'Many practitioners, scholars and students, both within and beyond Japan, mistakenly imagine human security to be something that applies to "others" in economically underdeveloped countries. Bacon and Hobson are therefore to be congratulated for refining and extending the human security approach, and demonstrating persuasively how it can be applied to the affairs of economically developed democracies in general, and more specifically to the 2011 "triple disaster", with which we are all still trying to come to terms here in Japan.'

Takashi Inoguchi, University of Tokyo, Japan

'The concept of human security has rarely been used or practiced within Japan, as if there were no vulnerable people in the country. Bacon and Hobson remind us that human security is relevant for all countries, and show how it can fruitfully be applied to the variety of challenges that Japan continues to face in the wake of the 2011 "triple disaster".'

Yasushi Katsuma, Waseda University, Japan

'This well-researched and cogently argued volume exposes the facade of the Japanese state›s commitment to Human Security in the light of the triple disasters of 3.11. It will be essential reading for anyone seeking to understand human insecurity in post-tsunami Japan.'

Giorgio Shani, International Christian University, Japan

'This book provocatively shows how the concept of human security applies to one of the most important recent disasters – the earthquake tsunami and nuclear disaster in Japan. It causes us – as students, academics or officials – to reflect humbly on how we can better protect human security, close to home, and further afield.'

Mihir Bhatt, Director, All India Disaster Mitigation Institute, India

Human Security and Japan's Triple Disaster

Japan has been one of the most important international sponsors of human security, yet the concept has hitherto not been considered relevant to the Japanese domestic context. This book applies the human security approach to the specific case of the earthquake, tsunami and nuclear accident that struck Japan on 11 March 2011, which has come to be known as Japan's 'triple disaster'. This left more than 15,000 people dead and was the most expensive natural disaster in recorded history.

The book identifies the many different forms of human insecurity that were produced or exacerbated within Japan by the triple disaster. Each chapter adds to the contemporary literature by identifying the vulnerability of Japanese social groups and communities, and examining how they collectively seek to prevent, respond to and recover from disaster. Emphasis is given to analysis of the more encouraging signs of human empowerment that have occurred. Contributors draw on a wide range of perspectives, from disciplines such as: disaster studies, environmental studies, gender studies, international relations, Japanese studies, philosophy and sociology.

In considering this Japanese case study in detail, the book demonstrates to researchers, postgraduate students, policy-makers and practitioners how the concept of human security can be practically applied at a policy level to the domestic affairs of developed countries, countering the tendency to regard human security as exclusively for developing states.

Paul Bacon is Associate Professor, School of International Liberal Studies, and Deputy Director of the European Union Institute, Waseda University, Japan.

Christopher Hobson is Assistant Professor, School of Political Science and Economics, Waseda University, Japan, and Visiting Research Fellow, United Nations University, Japan.

Routledge Humanitarian Studies Series

Series editors: Alex de Waal and Dorothea Hilhorst

The Routledge Humanitarian Studies series in collaboration with the International Humanitarian Studies Association (IHSA) takes a comprehensive approach to the growing field of expertise that is humanitarian studies. This field is concerned with humanitarian crises caused by natural disaster, conflict or political instability and deals with the study of how humanitarian crises evolve, how they affect people and their institutions and societies, and the responses they trigger.

We invite book proposals that address, amongst other topics, questions of aid delivery, institutional aspects of service provision, the dynamics of rebel wars, state building after war, the international architecture of peacekeeping, the ways in which ordinary people continue to make a living throughout crises, and the effect of crises on gender relations.

This interdisciplinary series draws on and is relevant to a range of disciplines, including development studies, international relations, international law, anthropology, peace and conflict studies, public health and migration studies.

Disaster, Conflict and Society in Crises
Everyday politics of crisis response
Edited by Dorothea Hilhorst

Human Security and Natural Disasters
Edited by Christopher Hobson, Paul Bacon and Robin Cameron

Human Security and Japan's Triple Disaster
Responding to the 2011 earthquake, tsunami and Fukushima nuclear crisis
Edited by Paul Bacon and Christopher Hobson

Human Security and Japan's Triple Disaster

Responding to the 2011 earthquake, tsunami and Fukushima nuclear crisis

Edited by Paul Bacon and Christopher Hobson

Routledge
Taylor & Francis Group
LONDON AND NEW YORK

First published 2014
by Routledge

Published 2016 by Routledge

2 Park Square, Milton Park, Abingdon, Oxon, OX14 4RN

and by Routledge
711 Third Avenue, New York, NY 10017

Routledge is an imprint of the Taylor & Francis Group, an informa business

First issued in paperback 2015

British Library Cataloguing in Publication Data
A catalogue record for this book is available from the British Library

Library of Congress Cataloging-in-Publication Data
A catalog record has been requested for this book

ISBN13: 978-1-138-01313-1 (hbk)
ISBN13: 978-1-318-64698-8 (pbk)

Typeset in Baskerville
by HWA Text and Data Management, London

Contents

Illustrations

Figures

Tables

Contributors

Paul Bacon is an Associate Professor of International Relations in the School of International Liberal Studies, Waseda University.

Andrew DeWit is a Professor in the School of Policy Studies, Rikkyo University.

Akiko Fukushima is a Senior Fellow at the Tokyo Foundation.

Christopher Hobson is an Assistant Professor in the School of Political Science and Economics, Waseda University.

Paul James is Director of the UN Global Compact Cities Programme and Professor of Globalization and Cultural Diversity at the University of Western Sydney.

Jeff Kingston is Director of Asian Studies, Temple University Japan.

Elizabeth Maly is a Senior Researcher, Disaster Reduction and Human Renovation Institution, Kobe.

Junko Otani is Director, East Asia Centre for Education and Research (Shanghai Office) and Associate Professor, Graduate School of Human Sciences, Osaka University.

Mayumi Sakamoto is a Designated Associate Professor at the Disaster Mitigation Research Center, Nagoya University.

Mai Sato is an Oxford-Howard League Post-Doctoral Fellow at the Centre for Criminology, University of Oxford.

Acknowledgements

The origins of this book lay in a workshop held in Tokyo in February 2012. This event was organized and hosted by the School of International Liberal Studies, Waseda University and the Institute for Sustainability and Peace, United Nations University (UNU-ISP), with the support of the Global Cities Research Institute, RMIT University. The workshop was funded by an 'Intellectual Exchange Conferences' Grant from The Japan Foundation. We would like to express our appreciation to these institutions, and especially The Japan Foundation, for making this project possible. At RMIT University, we would like to thank Paul James; at Waseda University we would like to thank Norimasa Morita and Naoko Sekimukai; and at UNU-ISP we would like to thank Yoshie Sawada, Mari Yamamoto and the interns who have assisted us at various points in this project.

The vast majority of this book was completed while Christopher Hobson was working at UNU-ISP and he is appreciative of the institutional support that was provided to him. We are especially grateful to Nicholas Turner from UNU-ISP, who has played a central role in shaping and assisting this project since the initial grant application. Nicholas might not be listed as a co-editor, but this book would not have been possible without his contribution. We would also like to express our gratitude to Khanam Virjee, Helen Bell, and their colleagues at Routledge. They have been helpful, professional, and a pleasure to deal with.

Both the editors are based in Japan, and were in Tokyo on 11 March 2011 when the earthquake and tsunami struck. The project was started by a desire to apply our knowledge to help understand what had happened, and this book represents a very small contribution towards Japan's efforts to recover from these terrible events. It builds on the work undertaken in the accompanying volume, *Human Security and Natural Disasters* (Routledge 2014), which provides the conceptual foundations for applying a human security approach to natural disasters, and examines a range of important case studies from around the world. In this volume, we consider the relevance of human security for Japan, and explore the many challenges and vulnerabilities the Japanese people have had to address in the years after 3/11. Together these books seek to clearly demonstrate that natural disasters are emerging as one of the greatest threats to human security in the twenty-first century, and that it is vital that we prepare better for them.

Paul Bacon and Christopher Hobson
Tokyo, January 2014

1 Human security comes home

Responding to Japan's triple disaster

Paul Bacon and Christopher Hobson

Taigan no kaji

Taigan no kaji is a Japanese expression that translates approximately as 'the fire across the river', and means that, even though a danger or problem can be seen, nothing is done about it because it is far enough away that it can be ignored or avoided: the people on *this* side of the river are safe. This kind of thinking has long been present in the way Japan has understood the human security doctrine. While the country has been one of its biggest international supporters, there has been an underlying assumption that human security is a concern *for others*. Yet one of the most important elements of the human security approach is the recognition that grave threats to the safety and well-being of people can be found everywhere. Just because Japan is an industrialized, wealthy democracy does not mean it is free of human insecurities. Certainly the kind of vulnerabilities people face will vary depending on the context, but human security's core concern with promoting 'freedom from fear' and 'freedom from want' is meant to have universal validity.

The relevance of the human security approach for Japan was demonstrated with horrible force on the afternoon of 11 March 2011, when a magnitude 9 earthquake occurred off the coast of Japan, triggering a massive tsunami that inundated the Tōhoku coastline. After the black waters of the tsunami had receded, left in its wake was a tragic scene of death and destruction that had left more than 15,800 dead, another 3,200 missing, whole communities destroyed and approximately $177 billion worth of damage (Japan Reconstruction Agency 2013, pp. 1–2). The tsunami also overwhelmed the inadequate defences of the Fukushima Dai-ichi nuclear power plant, resulting in a complete loss of power, with meltdowns subsequently occurring in three of its six reactors. This would ultimately become the second worst nuclear accident in human history, with 'Fukushima' now assuming a similar place to 'Chernobyl' in the global lexicon. As of November 2013, 1,605 people from Fukushima had died of indirect causes relating to the evacuation following the nuclear accident, and still more than 154,000 people remain evacuated from the region, with some having to abandon any hopes of returning to their homes (*Mainichi* 2013b; Japan Reconstruction Agency 2013, p. 15). More than three years later, the massive human, economic and environmental costs are still being counted, and Japan struggles with the difficult process of rebuilding.

In this book we argue that what has become known as Japan's 'triple disaster', comprising the earthquake, tsunami and nuclear accident, has forced a reconsideration of the way human security should be understood, and an acknowledgement of its applicability in the Japanese domestic context. After 3/11, human security was not just a concern for developing countries; it was not simply a problem 'across the river'. The purpose of this book is thus to use human security as a way of identifying the kind of vulnerabilities that were created and exacerbated following the triple disaster, as well as for considering how people have maintained their agency or been empowered during these difficult circumstances. In addition, almost all of the contributors live and work in Japan, and on a more basic level, it is a response to a tragedy by which we have all been directly affected. In looking at the specific case of the 2011 tsunami and nuclear accident, the book also contributes to the further development of the human security approach. It does so in two main ways. First, it is focusing on human security in the context of a stable, wealthy democracy, thereby countering the aforementioned tendency to regard it as only being of relevance for developing countries. Second, the book extends and refines the approach by applying it to a major natural disaster, which stands in contrast to the tendency to focus on insecurity generated by conflict or underdevelopment. In this regard, it builds on the work undertaken in the companion volume, *Human Security and Natural Disasters*, which provides a more extensive conceptual discussion on human security as well as a series of case studies (Hobson et al. 2014).

The introduction commences by outlining in more detail the human security approach and its relevance for considering natural disasters. It is suggested that the simple move of placing humans at the heart of our analysis is an important corrective, one that encourages a concern with the way vulnerabilities are constructed or reduced through social practices. After the human security approach has been introduced, the chapter will briefly apply it to the triple disasters, illustrating the different kinds of vulnerabilities that were created and exacerbated, while also identifying certain ways that people maintained their agency even in these incredibly difficult circumstances. The final section provides summaries of the chapters that follow.

Human security

The human security approach, as the name suggests, commences with a basic, but important assumption: that our primary focus should be directly on the safety and well-being of people. It is based on an 'expanded understanding of security where the protection and empowerment of people form the basis and the purpose of security' (UNSG 2010, p. 6). In itself this may seem rather self-evident, but it represents a valuable corrective to the longstanding habit of prioritizing state security. As the UNDP observed, for too long 'forgotten were the legitimate concerns of ordinary people who sought security in their daily lives' (UNDP 1994, p. 22). For most people, the greatest threats to their well-being do not come from war between states, but from problems such as hunger, disease, displacement, civil conflict and environmental degradation. This approach thus seeks to realign our understanding of security with the actual threats that most severely and most commonly impact people. It is in this sense that human security is meant to entail the 'protection from sudden and hurtful disruptions in the pattern of our daily lives' (UNDP 1994, p. 3). The ultimate goal is that people can be free from 'fear' and 'want', a situation in which people can live their lives with dignity.

Human security, as first set out in the 1994 UNDP Human Development Report, is meant to offer a holistic perspective, one that places equal weight on vulnerability caused by physical violence and that triggered by other factors, such as poverty, underdevelopment or natural disasters. Some have critiqued this as being overly broad, instead arguing that the remit of human security should be limited to a concern with physical violence. Yet doing this would largely defeat the purpose of adopting a human security approach in the first place. As noted, a fundamental premise underlying the doctrine is that vulnerability caused by hunger, disease or environmental damage can be just as detrimental to the safety and well-being of people as direct physical violence. In the case of natural disasters, in many ways the human consequences are not that dissimilar to those caused by war: widespread death, massive destruction, extensive displacement and heightened vulnerability. If the objective of human security is to reorientate 'our understanding of security towards incorporating the most severe and immediate threats that people face, it makes little sense to arbitrarily prioritise one kind over the other' (Hobson 2014a, p. 22).

In considering Japan's triple disaster, this volume is clearly adopting a 'broad' definition of human security, which corresponds to the way it has been understood in all the landmark texts: the 1994 UNDP Human Development Report, the 2003 Commission on Human Security report and the UN Secretary-General's reports of 2010 and 2012. Furthermore, these texts all explicitly identify natural disasters as posing a significant threat to human security. Indeed, the Commission on Human Security noted that natural disasters, along with economic crises and conflict, 'inflict the greatest shocks on society and people's human security' (Commission on Human Security 2003, p. 73). To date, however, those working with a 'narrow' definition have focused on threats emerging from war and conflict, while those that have adopted a 'broad' understanding have tended to look more at developmental issues. This book builds on the work undertaken in the accompanying volume, *Human Security and Natural Disasters*, which demonstrates in detail the theoretical and empirical ramifications of natural disasters for the human security agenda (Hobson et al. 2014). It does so by providing an in-depth analysis of a significant case study of human insecurity, one explicitly noted in the UN Secretary-General's most recent report on human security (UNSG 2012, p. 5). Given that extreme weather events are becoming more frequent as a result of climate change, Japan's experience may be a harbinger of what future disasters will look like.

One distinct advantage of the human security approach for understanding natural disasters is its emphasis on the direct and immediate role people play in creating, exacerbating or reducing vulnerability and risk. Natural disasters are never completely natural: human behaviours and social structures shape the way they unfold. In this regard, when considering the triple disaster there has been a tendency to draw a distinction between the natural disaster (earthquake and tsunami) and the man-made one (the nuclear accident). This separation is misleading, however, as it obscures the fact that the human element was present throughout. For the nuclear accident it is more obvious: the tsunami, a natural event, interacted with grossly inadequate preparations and poor decision-making to cause meltdowns in three of the reactors (NAIIC 2012). Yet this combination of human and natural elements was also present in the case of the tsunami: while it was unusually powerful, inadequate preparation also played an important part in what happened. Factors such as complacency over the risks posed by tsunamis, misplaced confidence in sea walls that proved to be not high enough and insufficient disaster training all contributed to the high death toll (Parker 2012). In this regard, a human security perspective accepts that it may be impossible to completely prevent natural hazards, but adequate preparation and planning can help limit the damage caused. This point is expressed by the Commission on Human Security: 'the infrastructure of protection may be imperfect, but it can help to counter threats, mitigate their force, support people threatened and create a more stable environment' (CHS 2003, p. 11).

In unpacking human security, it is valuable to return to the 1994 UNDP Human Development Report, which identifies four main elements of the approach:

- Human security is a *universal* concern.
- The components of human security are *interdependent*.
- Human security is *easier to ensure through early prevention* than later intervention.
- Human security is *people-centred*. (UNDP 1994, pp. 22–3)

In addition to these main elements, seven key areas of threat to human security are identified: economic security, food security, health security, environmental security, personal security, community security and political security. When combined, these four core components and seven major threats provide a useful framework for understanding how human security is threatened, protected or reduced. This is covered in detail in the accompanying volume (see especially Bacon and Hobson 2014, pp. 4–13), but it is helpful to recap some of the main points here.

Universal

According to the 1994 Human Development Report, 'human security is relevant to people everywhere, in rich nations and in poor' (UNDP 1994, p. 3). Certainly the nature and extent of insecurity will vary depending on the context, but all people are liable to experience vulnerability. This observation was reaffirmed by the UN Secretary-General: 'people throughout the world, in developing and developed countries alike, live under varied conditions of insecurity' (UNSG 2012, p. 5). This insecurity is reinforced by the increasingly rapid onset of climate change, which is resulting in more frequent and extreme weather events, as well as making existing infrastructure increasingly obsolete for dealing with these risks. Developing states are certainly much more vulnerable, but as Japan unfortunately discovered, even a highly industrialized and stable democracy can be profoundly affected. Indeed, the assumption that a country like Japan is essentially devoid of human security threats can lead to such issues being overlooked. For instance, the media narrative of affected Japanese maintaining a 'quiet dignity' in the face of considerable hardships after the tsunami (*Daily Mail* 2011) obscured cases of violence, rape, looting and theft that took place (Goldenberg 2011; Polaris Project 2011). These problems were certainly not as widespread as in some other post-disaster contexts, but it is important to recognize that these threats to personal security could also be found in Tōhoku.

Interdependent

The human security approach emerged in the 1990s, when globalization was the *zeitgeist* (Rosenberg 2005). In that context, the reference to interdependence in the 1994 document simply meant that human security threats do not respect state boundaries. Neither the Commission on Human Security, nor the UN Secretary-General's reports on human security greatly elaborated on this aspect of the approach. Yet as we argue in the accompanying volume, 'interconnectedness is not only about threats combining and interacting across global borders, it is also about how they compound within a specific local context' (Bacon and Hobson 2014, p. 6). Indeed, one of the greatest potential contributions of the human security approach is providing a lens through which to examine the way different forms of vulnerability interact. Utilizing the different categories of human security threat it is possible to 'look for "threat multipliers" – where insecurities are reinforced – and "solution multipliers" – where it may be possible to address more than one threat at the same time' (Bacon and Hobson 2014, pp. 6–7). This is particularly relevant for natural disasters, in which a myriad of threats appear and interrelate. Examples of the interdependent nature of threats will be provided below when examining the seven main categories of threat to human security.

Prevention

Human security is easier to ensure through early prevention than later intervention: the most effective way of addressing a threat is dealing with it beforehand. The Hyogo Framework for Action, a document that exemplifies what a human security approach should look like, conveys this point well: 'the promotion of a culture of prevention, including through the mobilization of adequate resources for disaster risk reduction, is an investment for the future with substantial returns' (UNISDR 2007, p. 5). This concern with prevention leads to an emphasis on the role people can play in actively reducing their vulnerability.

Prevention is relevant to disaster risk reduction in general, and the Japanese triple disaster in particular, in at least three significant ways: in mitigating the immediate physical impact of earthquakes and tsunamis on people and infrastructure; in provoking more transparent and comprehensive debate on the impact of earthquakes and tsunamis on nuclear plants, and the risks of nuclear energy in general; and in creating better social policy for populations affected by disaster in the response and rebuilding phases.

The importance of prevention was made particularly clear by the nuclear accident. After the triple disaster, a government-appointed panel of experts determined that the Tokyo Electric Power Company (TEPCO) was ill-prepared for a crisis and that its workers and their managers were inadequately trained to cope with an emergency situation (Hatamura 2012). The Kurokawa report was even more damning:

> There were many opportunities for taking preventive measures prior to March 11. The accident occurred because TEPCO did not take these measures, and NISA [the Nuclear and Industrial Safety Agency] and the Nuclear Safety Commission (NSC) went along.
>
> (NAIIC 2012, p. 16)

These issues are explored in depth in Jeff Kingston's chapter, where he demonstrates how a failure to properly prepare created a situation of extreme human insecurity. The terrible experience of the Fukushima accident has also led the Japanese public to turn against nuclear power, as they have become sceptical over claims that it is possible to adequately prepare for future earthquakes and tsunamis. As Bacon and Sato consider in their contribution, public sentiment increasingly supports the argument that the best way to prevent a nuclear accident is by abandoning nuclear power completely.

People-centred

On a basic level, stating that human security is 'people-centred' reflects the underlying assumption that security should be defined directly in relation to the well-being of people. This may seem like an obvious injunction, but too often it is not followed. TEPCO's behaviour offers a particularly egregious example of this. As the reactors at Dai-ichi went into meltdown, TEPCO tried to delay for as long as possible the use of seawater to cool the reactors, as this would render them unusable in the future (Maeda 2012). And in the months that followed, TEPCO avoided dealing with the problem of contaminated groundwater at the plant because of concerns over how the stockmarket would assess this considerable financial liability (*Asahi Shimbun* 2013b). In situations such as these, TEPCO repeatedly prioritized its own interests ahead of the people that had been affected, and the government failed to regulate it to do otherwise. It was not only TEPCO who behaved in this manner, however. As Kingston details in his chapter, in the months after the disaster many politicians were more interested in forcing Naoto Kan out of power than in dealing with far more significant issues to do with assisting the affected areas. In such contexts, human security gives 'a different – and valuable – starting point for considering decisions, one that offers a perspective more attuned to the everyday concerns of people' (Hobson 2014a, p. 25).

There is a further aspect to human security's people-centredness, one which has often been lost as a result of casually eliding 'people' with 'individuals'. The use of the former term in the 1994 UNDP report was not accidental; it reflects that human security 'is concerned with how people live and breathe in a society' (UNDP 1994, pp. 22–3). Security and insecurity are social phenomena, conditions generated through people interacting with each other. As such, a human security approach calls for examining the way social structures can increase or reduce vulnerability. This is something explored by Junko Otani, Elizabeth Maly and Mayumi Sakamoto in their respective chapters, who all emphasize the importance of community bonds during the difficult rebuilding process, and the necessity of recovery plans that maintain and foster social interaction. In her contribution, Akiko Fukushima makes a related argument, by considering how shared culture can be an important resource for dealing with the traumatic memories of the disaster.

Threats to human security

These core components of human security – universality, interdependence, prevention and people-centredness – offer a useful way of understanding the 'humanness' of natural disasters. They provide a framework for considering how human vulnerabilities are created and exacerbated, how they interact, and how – ideally – they can be reduced and people's agency can be supported and strengthened (see also Hobson 2014a). This can be further seen through considering the seven main forms of human security threats: economic security, food security, health security, environmental security, personal security, community security and political security. As a way of illustrating the full range of vulnerabilities the triple disasters gave rise to, the chapter will now work through these seven areas of threat in reference to the 2011 Tōhoku earthquake, tsunami and nuclear accident.

Economic security

When natural disasters strike, people are not impacted equally. In most situations, it is the more vulnerable parts of the population that suffer disproportionately. This has certainly been the case in Tōhoku. As Hobson notes in his chapter, the region was already suffering from problems of economic and community insecurity *before* the disaster as a result of depopulation and a shrinking economy. These trends have been considerably exacerbated by the tsunami and nuclear accident. The economic toll has been considerable. Many people in the region have lost their livelihoods: businesses have closed down, companies have reduced staff or relocated, self-employed people have lost their customers and those working for daily wages have found it much more difficult to find new employment. The farming and fishing industries have been particularly hard hit as a result of the nuclear accident. Once one of Japan's 'breadbaskets' famous for its fresh produce and fish, now goods with the 'Fukushima' label are avoided because of radiation fears. These economic hardships have left affected people much more vulnerable to any future shocks, be it another natural disaster or other events such as an economic downturn.

The economic costs of the triple disaster go well beyond the immediate Tōhoku region. According to the Japanese Reconstruction Agency (2013) the economic damage caused by the tsunami was approximately $177 billion. Meanwhile, estimates for costs stemming from the nuclear accident continue to rise, with decontamination and compensation expected to be more than $400 billion, and more than another $100 billion required for decommissioning Fukushima Daiichi (Saito et al. 2013). While TEPCO is nominally responsible, it is becoming increasingly clear that the state will have to cover much of the cost (Hamada 2013). This is hardly an insignificant burden considering that Japan is second only to Greece in terms of government net debt, equal to 140 per cent of its GDP (Alexander 2013). And as the Greece case has sadly demonstrated, economic insecurity leads to increasing community and political insecurity. If a similar kind of crisis unfolded in Japan, however, the consequences would be global in scope, given that Japan remains the world's third largest economy. This reflects the interconnected nature of human security threats, both within a domestic context and internationally.

Food security

Food security was primarily a concern in the immediate period after the tsunami struck. There were problems with getting supplies to the affected areas because planning had been based on the experiences of the 1995 Kobe earthquake, and authorities were not prepared for dealing with the combined challenges of a tsunami and a nuclear accident. For the first weeks after 11 March, people in shelters were given mainly rice balls and bread, which are not very nutritious. Representative of what they had to endure, one survivor explained at the time: 'I'm hungry. So are my four children. All we have to eat each day is one banana and a rice ball' (Watts 2011). Even in a place like Japan, a wealthy country with excellent infrastructure and transportation, there was serious food insecurity immediately afterwards due to problems with getting nutritious food and clean water to those that needed it. This throws into stark relief the difficulties that lesser developed countries must deal with.

The nuclear accident also posed a set of unique challenges in terms of food security. During the crisis there was a major concern with preventing contaminated food and water from entering into the food chain. Shortly after the accident Japan's Ministry of Health, Labour and Welfare began testing food and water for radioactive contaminants, and also restricted the distribution of potentially contaminated foods. Yet many questioned whether enough was being done. At the time Mariko Sano, Secretary-General of a women's organization in Tokyo, complained that 'the government is so slow to move. They've done little to ensure food safety' (Takada 2011). Lack of confidence in the government, as well as fears that the accident would worsen, led people across Japan to hoard food and water, which actually further contributed to food shortages in the affected area. More than three years after the disaster, fears about food safety linger.

As of the time of writing, the Ministry continues daily tests for levels of radioactive contaminants in food from the affected prefectures (Japan Ministry of Health, Labour and Welfare 2013). It is not enough to convince many people who still question pronouncements that food from Fukushima is safe, however. This is particularly the case with young mothers, concerned about the impact radiation might have on their children. Here is another example of the interconnected nature of security threats: fears over radiation have had massive economic consequences for the Fukushima farming and fishing industries, leaving farmers and fishermen more vulnerable. At the same time, fears over radiation have led to increased stress levels for mothers, which is directly relevant for the next category of health security.

Health security

As with food security, the most immediate health threats were in the period directly after 11 March. On the whole, the Japanese government and aid agencies such as the Red Cross were very effective in their response. Inadequate preparations for a nuclear accident, combined with poor risk assessment, did result in some serious failures, however. In one case, a botched evacuation of Futaba Hospital, located 4.6 kilometres from Dai-ichi, led to 130 patients being stranded for up to five days with limited food and medical supplies, resulting in nineteen deaths (Otsuki 2012). Not only does this reflect a lack of preparation for a nuclear emergency, it also illustrates the considerable challenges that arise from managing evacuations in which a high number of the affected people were elderly. Considering it has the fastest ageing population in the world, Japan needs to adequately prepare for caring for elderly victims in the case of future disasters.

For those that managed to survive, anger, anxiety, depression, fear, hopelessness and survivor's guilt are some of the feelings that they must deal with. At the most extreme, there have been sad stories of people committing suicide because of the trauma associated with losing their homes and businesses after the nuclear accident. There has also been a notable increase in alcohol abuse in the affected region, with one study indicating that the rise has been particularly pronounced in coastal cities hardest hit by the tsunami (*Mainichi* 2013c). These mental health problems are made more challenging by Tōhoku being a traditional part of Japan where it is not normal to openly discuss one's feelings, especially with strangers. In dealing with this challenge, avoiding one-to-one counselling in favour of 'coffee houses' where evacuees can interact with one another in a group environment is proving to be more successful (Ellis 2013). This is where human security's concern with 'people' and not 'individuals' is especially important – mental health challenges must be managed in a way that takes into account local social norms and customs.

Environmental security

The human consequences of the tsunami were so severe there is a tendency to overlook the impact it had on nature. The tsunami generated more than 20 million tons of debris, of which 5 million tons was washed out to sea: 3.5

million tons of this debris was deposited on the seabed along Japan's coast and much of the remaining 1.5 million tons is still drifting in the ocean (Kantei n.d.). Much of Tōhoku's famous picturesque coastline was obliterated, notably its once famous pines along the shores. The tsunami also caused farmland in the region to become heavily salinated, which in turn has had consequences for food security, as subsistence rice farmers were particularly hard hit (Plett 2012). The interconnected nature of security threats, and the trade-offs that are sometimes necessary, can also be seen in a concern with preventing future tsunamis leading to the building of 'stunningly tall and wide seawalls' that risk destroying important ecosystems along the coastline (Bird 2013). In this case, one core element of human security – a concern with prevention – is at odds with another – environmental security. There is a need to consider how it is possible to harmonize the two, and explore measures that will protect both people and natural ecosystems.

As with the tsunami there has been a tendency to focus on the human costs of the nuclear accident, but the environmental consequences have also been far-reaching. Areas northwest of Dai-ichi were most heavily contaminated with radioactive materials, with a majority of the contaminated area being forest, which will be particularly difficult to treat without damaging the ecosystems. Decontamination efforts have been strongly criticized for being costly and ineffectual. The Japanese government has already awarded $13 billion in contracts mainly to construction companies with limited experience in decontamination, which have also relied on an exploitative system of subcontracting, where untrained casual labourers are tasked with the most difficult and dangerous jobs (Tabuchi 2012). This is another example of the interconnected nature of human security threats, where the environmental problem of decontamination is leading to vulnerable people being exploited. Given that these attempts at decontamination have been largely ineffectual, from a human security perspective the money would be much better used for resettlement grants or compensation to disaster victims, instead of awarding it to major corporations. These decontamination efforts, which have involved removing topsoil, turf stripping and the clipping and removal of leaf waste, have also given rise to a large amount of contaminated waste that authorities have struggled to store. In one city in Fukushima, bags of contaminated waste were left unprotected next to public apartment complexes and parks, in turn creating health risks for nearby residents who had not been made aware of the risks (*Mainichi* 2013a). Determining what to do with the large amount of contaminated waste, along with finding ways to manage ongoing issues with contaminated groundwater at the Dai-ichi plant, are major practical and political problems to which the government has yet to find long-term solutions.

Personal security

At 2:46pm on Friday 11 March 2011 hundreds of thousands of people went from going about their daily lives to being plunged into extreme insecurity. According to the UN Secretary-General's 2010 report, 'the concept of human security acknowledges that due to catastrophic events people may be faced with sudden insecurities and deprivations' (UNSG 2010, p. 7). This description is certainly appropriate to describe the situation after the tsunami struck the Tōhoku coast. A week later and a total of 386,739 people had been evacuated to more than 2,000 evacuation centres (Hasegawa 2013, p. 15), in which they were forced to survive in difficult circumstances, with little privacy and few comforts. And, as Otani details in her chapter, conditions were more challenging for groups of people that were already vulnerable, such as the elderly or disabled. Even in these difficult circumstances, by emphasizing the human and social element it was possible to increase the sense of security and safety felt by evacuees. Small but culturally important acts, such as taking off shoes when in the shelter and providing baths, 'started a transition from a state of apathy, memory loss, high tension and anxiety to a process of re-creating social order and stability' (Steger 2012).

Following the tsunami and nuclear accident there may not have been widespread scenes of violence or lawlessness, but this does not mean that there were not problems of violence and discrimination. In this regard, looking at gender relations is a useful way of quickly identifying some of the most severe and immediate human security threats that exist in a certain situation (Hobson 2014b, pp. 184–5). Unfortunately women evacuees suffered many of the same kinds of vulnerabilities and problems found in other post-disaster contexts. One recent study identified a range of forms of abuse that took place, including:

> quid pro quo type abuse (e.g., demanding sex in exchange for food or shelter); exploitative, coercive abuse of a sexual nature by disaster response personnel and volunteers; physical, sexual, and/or verbal assaults against women by acquaintances; and sexual abuse against boys and girls.
>
> (ICESCR Gender Report Project 2013, p. 25)

The vast majority of abuses were committed by someone the victim knew; and in some cases by community leaders or disaster responders. There are clear signs that such problems have persisted, with domestic violence rising significantly in Tōhoku after the disaster, and women survivors also having experienced 'physical, sexual, psychological, verbal and economic violence' (Yamashita 2013, p. 2). These are the most extreme cases of environments that lacked sensitivity to gender and sexuality issues. In the Fukushima, Iwate and Miyagi prefectures, approximately 97 per cent of the community leaders in charge of designing and operating the evacuation centres were men, which unsurprisingly led to a lack of awareness of women's needs (Gender Equality Bureau 2012, p. 4). Even though Japan is fortunately free of the violence and physical insecurity many other places suffer from, these examples point to different ways in which people's personal security was impacted by the disaster.

Community security

The significance of gender relations for shaping personal security connects it to the category of community security. In this regard, human security commences from the basic observation that we do not live in isolation – our lives are given meaning by our ties to family, friends and the societies in which we live. These relationships can be destroyed, broken or severely tested following a natural disaster. After the tsunami, members of small villages were not always placed in the same shelters or temporary housing, which took away vital support networks for people. In their chapters, Maly and Sakamoto both emphasize the importance of promoting 'life recovery', an approach that seeks to foster the social bonds that help people deal with the difficult rebuilding process. This is particularly challenging in the Fukushima region. As noted, it was already suffering from a shrinking population, a trend that has sharply increased as many families and younger people choose not to return. As a result, towns are being recomposed of mainly elderly citizens, raising questions about their long-term viability, an issue Maly considers in her contribution.

The triple disaster has also placed considerable strain on evacuated families, who are forced to live in small, uncomfortable temporary housing with little private space. Even three years later many families still do not know when they can build their new homes, with the number of applicants greater than the amount of land available in some areas. As one evacuee put it: 'we only want to secure a place to live' (*Asahi Shimbun* 2013a). If one recalls that human security is about sustaining the conditions of daily life, this quote conveys the heart of the problem. These difficult conditions are even more pronounced in regards to the nuclear accident. A common situation is that the husband remains in Fukushima to work, while the wife and children move to another part of Japan where there are no concerns about radiation. This places huge emotional strains and economic burdens on families, and has given rise to conflicts in situations where partners hold different views of how dangerous the radiation threat is. There is even a term – *genpatsu rikon* ('atomic divorce') – to describe marriages that have broken down due to stresses related to the nuclear accident.

Political security

When it comes to political security, as with personal security, Japan is a relatively well governed and stable country free of widespread human rights violations, and people's rights were largely respected after the triple disaster. Yet there have been problems, especially with transparency and how information has been shared. The way the nuclear accident was handled has seriously shaken people's confidence and trust in the Japanese state. As the accident at the Fukushima Dai-ichi nuclear plant unfolded, the government and TEPCO failed to behave in an open manner and provide honest, accurate information about what was happening, and what might happen. A string of revelations and scandals relating to the Fukushima accident, combined with the difficult reconstruction process, has left many people increasingly sceptical and frustrated with Japanese elites. According to one survey, there was a 'precipitous drop in trust in Japan', with significant decreases in levels of trust in the government, the media, NGOs and business (Edelman 2012, p. 5). Meanwhile, Japan dropped from 22nd to 53rd place on the 2013 Press Freedom Index because of 'a lack of transparency and almost zero respect for access to information on subjects directly or indirectly related to Fukushima' (Reporters Without Borders 2013, p. 3). These concerns have been reinforced by the introduction of an oblique and wide-ranging secrecy law in December 2013, which was quickly forced through parliament with little debate, and could be used to restrict information in future situations similar to Fukushima. Withholding information in this manner directly contradicts human security's concern with 'developing the capabilities of individuals and communities to make informed choices' (Commission on Human Security 2003, p. 4).

What this discussion clearly points towards is the relevance of the human security approach for Japan. Human security is not simply a problem 'across the river', nor did it suddenly become relevant for the country on 11 March 2011. The various chapters in this volume demonstrate in different ways the kinds of insecurities that have arisen as a result of the triple disaster, and what needs to be done to address them.

Chapter outlines

Building on the introduction, Paul Bacon illustrates how Japan developed its own interpretation of human security, and has attempted to use it as a device to achieve Japanese foreign policy and development-assistance objectives. He shows how this approach has attracted strong criticism within UN circles. Bacon argues that Japan has been unfaithful to the 'philosophy' of human security, as espoused in the 1994 UNDP report, committing itself to a diluted version of the concept that does not reflect the claim and demand that human security be preventive, people-centred, universal and interdependent. Bacon examines how, when and why the Japanese conception emerged in its current politicized and exclusionary form, and in so doing, identifying the human error which led to the establishment of a conception that removed the Japanese themselves from the legitimate purview of human security. The overall conclusion offered is that the Japanese approach has not adequately provided for the achievement of human security within or beyond Japan's borders, and that it is time for a more inclusive rethink of the way in which the Japanese government conceives of human security.

In the following two chapters, Jeff Kingston and Christopher Hobson respectively investigate the political and social consequences of failing to adopt a human security approach towards disaster preparation, response and recovery. In his blistering critique, Kingston notes that, although the nuclear accident was precipitated by a natural hazard, it was first and foremost a man-made disaster. He argues that poor risk management, a failure to comprehend the lessons of previous disasters, along with institutionalized complacency about risk, were major factors increasing the likelihood of both a major accident and a fumbling crisis response. There were institutionalized failures to prepare for worst-case scenarios and to nurture a culture of safety where this should have been a priority. He documents how, as the tragedy unfolded, the key institutions and actors failed to communicate with each other and the emergency response was uncoordinated and ad hoc. This grave series of failures is all the more alarming because it happened in Japan, a nation known for its meticulous disaster emergency planning. This reinforces the fact that human security concerns are relevant within wealthy stable democracies.

For Kingston, the emergency response to the Fukushima nuclear accident was grossly inadequate by any measure. There had been no evacuation drills or disaster training because TEPCO had not wanted to alarm nearby residents. Kingston claims that the government did not intervene to insist on such drills because it was complicit in perpetuating the myth of 100 per cent safety that the nuclear industry had propagated to promote the expansion of nuclear power. Kingston identifies that this has spawned an anti-nuclear energy movement driven by newfound citizen awareness about how the government and utilities were poorly prepared to cope with a disaster that should have been a priority concern. The Japanese people, Kingston concludes, now understand only too well that their human security was compromised to accommodate nuclear industry interests.

In his chapter, Christopher Hobson considers the human security of the workers at Fukushima Dai-ichi, 'hidden' victims of the triple disaster. Understandably everyone at Dai-ichi from senior management down to the workers at the plant is seen as collectively responsible for the catastrophe. For Hobson, this perception ignores that the vast majority were contract labourers pushed into this line of work out of economic necessity. In the language of human security, there were not only problems of 'community security', but also 'economic security' and 'health security'. He describes how the nuclear industry has been built on a highly exploitative structure that takes advantage of the economically weak position of its employees, and has exposed them to considerable health and safety risks. These vulnerabilities were reinforced by the disaster. Many workers lost family, friends and colleagues, had their homes destroyed and were forced to evacuate. While suffering from these experiences they were forced to work at Dai-ichi as three of its reactors went into meltdown. In this sense, they were 'double victims', suffering very directly from both the tsunami and the nuclear accident. Hobson then considers the impact these conditions have had on these workers, focusing on the serious mental health issues that have arisen. The workers at Fukushima suffered greatly in the aftermath of the nuclear accident, and Hobson suggests that there is a need for their insecurities to be better recognized and addressed. Not only is the well-being of these workers a moral concern, it is also a very practical one, as they are absolutely crucial for successfully decommissioning the Dai-ichi plant.

In his chapter, Paul James argues that human security should have the concept of 'the human' embedded at the heart of it. Human security should not be a vague add-on to harder edged areas of security such as military security or state security. Nor should it be reduced to a few complementary factors such as food security, economic security and personal security, linked by the notion of 'freedom from want'. Instead, human security should become the centre of all discussions of security, and the basis of thinking about all security – including military security and disaster management. With James's alternative framing, 'the human' or 'the social' have to be treated as the foundational category of life. James argues that we need to redefine 'security' and 'insecurity' so these concepts no longer rest on the modern liberal absolutes of 'freedom from want/ freedom from fear'. Rather, human security should be understood as being grounded in processes of social sustainability and resilience in which security is something relational. James concludes by noting that this new interpretation of human security includes, first, the sustainable protection and provision of the material conditions for meeting the embodied needs of people, and secondly the protection of the variable existential conditions for maintaining a dignified life. Within this definition the core focus of human security should be on empowering those most vulnerable.

James's thoughtful reflection on the centrality of the social dimension of human security, and the need to emphasize resilience and empowerment, provides a platform for the four subsequent chapters from Mayumi Sakamoto, Elizabeth Maly, Junko Otani and Akiko Fukushima. Each identifies the central importance of community security as Japan recovers from the triple disaster, illustrating in a range of ways how it works as an important 'solution multiplier', having positive consequences for other dimensions of human security, such as personal and health security.

Sakamoto's chapter discusses human security in the disaster response and recovery phases, drawing on experiences from both the 1995 Great Hanshin-Awaji Earthquake and the 2011 Great East Japan Earthquake. She argues that insufficient attention has been paid to post-disaster recovery, which has led to phenomena such as 'disaster-related death' and 'dying alone', and also to inadequate provision for those with disaster-related disabilities. In this context, Sakamoto explains how the concept of 'life recovery' was one of the lessons derived from the Hanshin-Awaji Earthquake, and considers how it is being applied following the 2011 triple disaster. This approach emphasizes the interrelatedness of several different factors such as housing, social ties, community, health, preparedness, economy and governance. The life recovery approach closely reflects the people-centred dimension of human security, which is concerned not simply with the material needs of individuals, but also with how security is created and maintained through social ties. It also reflects the priority assigned to prevention, and the interrelatedness of threats to human security. Recovery policy should thus recognize that in functioning societies people are embedded in social networks, and build the relevant safety mechanisms into social practices from the beginning. Recovery policies that do not recognize the significance of community security and do not make this the cornerstone of their strategy will likely fail.

Elizabeth Maly's chapter considers what is needed for a people-centred approach to housing recovery in Tōhoku. Maly argues that the plight of evacuees who have fled Fukushima after the nuclear accident raises serious concerns related to human security. A people-centred housing recovery for Fukushima evacuees requires special and immediate attention to unique housing and planning issues. Housing recovery in Fukushima shares some issues with the disaster area as a whole, such as the needs of vulnerable populations and ageing communities, and balancing risk-based planning with other needs for housing, livelihood and sustainable communities. However, the unprecedented scale and complexity of the earthquake mean that the post-disaster recovery will be much more challenging. To date, the actual housing recovery situations vary greatly. There are some examples of excellent initiatives that support people-centred approaches to housing recovery, such as the provision of social support staff, community spaces and activities in temporary housing; attempts to keep communities together; high-quality/wooden temporary housing; and planned small-scale public housing. There are also many examples of the same mistakes from Kobe being repeated in Tōhoku, such as poor-quality temporary housing and scattered communities. In addition there are new and massive challenges for housing recovery in Tōhoku, such as residents' facing displacement and relocation because their homes are contaminated by radiation, and a disaster area that was already facing regional economic decline and depopulation. In conclusion, Maly notes, although Japanese policies have evolved towards a more people-centred approach to housing recovery, there are significant and urgent changes that must still be made.

Junko Otani demonstrates the interconnected nature of human security threats by examining the impact of natural disasters on the health of older people, looking primarily at the Kobe and Tōhoku disasters. Otani explains that in the 1995 Great Hanshin-Awaji Earthquake more than half of those who died were aged over 60. With regard to the 2011 disaster, the elderly were again one of the most vulnerable groups, with 92 per cent of deaths due to drowning, and 65 per cent of those killed aged over 60. She shows how in these disasters it was not the immediate effects that disproportionately affected older people, but a number of secondary consequences. The elderly that do survive are confronted with difficult conditions at both evacuation centres and in their homes. With regard to evacuation centres, Otani argues that it is important not only to build structures which are earthquake resilient, but also in a way that takes account of the needs of vulnerable groups, such as the elderly and the disabled. Evidence from the Kobe case suggests that, after one year, the kind of support required changed. Those who remained needed more help, as many of them were marginalized people, including those living on welfare, with alcohol dependency or those with no family. Temporary shelters built to last for one year were occupied for more than three years. This experience suggests that when building temporary housing special attention needs to be paid to the most vulnerable parts of the community. In this regard, community security plays a key role in supporting health security in the long term, especially for older people.

In her chapter, Akiko Fukushima argues that, while culture has been recognized as an important dimension of human security, its role has not been sufficiently emphasized or explored. She attempts to address this gap in the human security literature by looking at some of the cultural issues which are relevant in the context of natural disasters through considering the Japanese case. Fukushima argues that cultural activities can play a highly significant role in the psychological rehabilitation of people and communities in the late-response and recovery phases following natural disasters. She explains how initially, right after the disaster, the mood in Japan was such that, out of respect, cultural activities were cancelled, but this proved short-lived. Fukushima gives concrete examples of how artists rushed to Tōhoku to support survivors through cultural activities; actors, singers, musicians and sportsmen, some originally from affected areas, were among the first to go to the devastated areas. Survivors also took their own cultural initiatives. This matches closely with human security's concern for people-centred grassroots action. Fukushima also gives examples of cultural activities contributing to the restoration of community security, describing three local festivals, one region-wide festival, the successful re-establishment of a culturally significant and economically viable pottery-making practice, and a museum exhibition.

Fukushima argues that cultural activities have had a significant psychosocial impact on survivors' lives in post-disaster settings: they turn despair into hope for the future, heal the psyches of survivors, rebuilding the self-respect and pride lost as a result of the disaster, and empower local populations. Disaster victims require adequate attention to be paid to their trauma, if their suffering is not to be long-lasting or lifelong. For Fukushima, this is where cultural activities come into their own, as a medium to help heal psychological scars resulting from the loss of family and friends. Fukushima concludes by noting that the triple disaster endangered human security, and that a culturally-informed, people-oriented human security approach is required, if local people are to be empowered, and post-disaster reconstruction is to fully succeed.

The final two chapters look to the future. Bacon and Sato take up the nuclear issue where Kingston finished, bringing a human security lens to bear on the energy mix question in post-Fukushima Japan. They focus in particular on two of the four key elements of human security: prevention and people-centredness. Bacon and Sato conclude that the risks and threats to human security engendered by the use of nuclear energy outweigh any benefits that could reasonably be argued to accrue from its use. The notion of prevention performs a further 'trumping' function, emphasizing and placing a premium on the downside risk of the use of nuclear energy.

Bacon and Sato also propose that the Abe administration's energy strategy is not sufficiently people-centred, as it goes against the wishes of the Japanese public. They argue that, although the LDP won the elections of December 2012 and July 2013, this does not necessarily mean that Abe received a mandate for his nuclear strategy. Bacon and Sato demonstrate how, in the three years since the Fukushima crisis first manifested, the Japanese public has consistently shown that it is opposed to nuclear energy in a series of opinion polls, and through protests and demonstrations. They also focus in some detail on a deliberative poll commissioned by the DPJ, through which representatives of the Japanese public also expressed clear opposition to nuclear power. A genuinely people-centred approach should acknowledge the resilience of this public opposition to nuclear power, and this sentiment should be incorporated into the larger debate on the energy mix question in Japan.

Japan's Self-Defence Force played a leading role in responding to the triple disaster. Its successful collaboration with US forces, 'Operation Tomodachi', was another positive. These kinds of issues are considered in Andrew DeWit's chapter, which explores how the military role in foreign humanitarian assistance and disaster response (HADR) has been transformed by accelerating climate change. DeWit argues that it is imperative to expand the institutionalization of military-centred HADR collaboration. The collaborative frameworks that must be bolstered are, firstly, those between military and civil-society HADR networks (such as NGOs) and secondly, those among the American and other global militaries. DeWit discusses several reasons for doing this, but the main reason is that the military increasingly possesses the means needed for prompt and effective HADR intervention, and that makes it a necessary part of any equation for adapting to the ravages of climate change.

DeWit notes that between 2004 and 2010 the American military alone was deployed at least 40 times to support HADR operations. The US military deployment to Japan in 2011, during 12 March to 4 May 2011, involved 24,000 US service members, 189 aircraft, 24 naval ships and cost US$90 million. In spite of these developments, many actors within the expanding network of HADR efforts, as well as analytical observers of military affairs generally, continue to see the military role in HADR as exceptional or undesirable. Nevertheless, DeWit believes it is fruitful to investigate how military resources are being deployed in HADR and how that might be increased. He stresses that military forces need to be further incentivized to collaborate on HADR, as much and as rapidly as possible, in order to steer them away from the mounting risk of direct confrontation. He believes that the evidence suggests that it might be possible to transform military priorities in the face of new security and humanitarian challenges. This perspective is in keeping with that of the other chapters in the book, which emphasize that the military component is but one dimension of human security, when security is viewed in a more holistic manner.

The volume ends with a short conclusion by one of the editors, Christopher Hobson. Whereas this introduction has focused primarily on vulnerabilities created and exacerbated by the triple disaster, the final chapter seeks to explore the empowerment component of human security. As well as considering some of the 'positive' developments that have occurred, the chapter reflects on how the disaster helps us understand the importance of maintaining daily life as part of a human security approach.

References

Alexander, D. 2013, 'The World's Largest Debtor Governments, 2013', *Forbes*, 11 Aug. http://www.forbes.com/sites/danalexander/2013/11/08/worlds-largest-debtor-governments-2013, viewed Dec. 2013.

Asahi Shimbun 2013a, '2011 disaster survivors stuck in housing limbo', 11 Sept., https://ajw.asahi.com/article/0311disaster/life_and_death/AJ201309110073, viewed Dec. 2013.

Asahi Shimbun 2013b, 'TEPCO delayed tackling radioactive water in 2011 due to bankruptcy fears', 18 Sept., http://ajw.asahi.com/article/0311disaster/fukushima/AJ201309180063, viewed Dec. 2013.

Bacon, P. and Hobson, C. 2014, 'Incorporating natural disasters into the human security agenda', in C. Hobson, P. Bacon and R. Cameron (eds), *Human Security and Natural Disasters*, London: Routledge, pp. 1–21.

Bird, W. 2013, 'Post-tsunami Japan's push to rebuild coast in concrete', *Asia-Pacific Journal*, 11/21(1), 26 May, http://www.japanfocus.org/-Winifred-Bird/3945, viewed Dec. 2013.

Commission on Human Security 2003, *Human Security Now*, New York: CHS, http://isp.unu.edu/research/human-security/files/chs-security-may03.pdf, viewed Aug. 2013.

Daily Mail 2011, 'No looting, just keeping calm and carrying on: Ordinary Japanese try to get back to normality', 15 March, http://www.dailymail.co.uk/news/article-1366532/Japan-tsunami-earthquake-Keeping-calm-carrying-normality.html, viewed Dec. 2013.

Edelman 2012, *2012 Edelman Trust Barometer: Global Results*, 19 Jan., http://trust.edelman.com/trust-download/global-results, viewed Dec. 2013.

Ellis, E. 2013, 'Getting on with life in Japan after tsunami's mental challenges', *Vancouver Sun*, 21 Dec., http://www.vancouversun.com/news/Getting+with+life+Japan+after+tsunami+mental+challenges/9312299/story.html, viewed Dec. 2013.

Gender Equality Bureau 2012, *Disaster Prevention and Reconstruction from a Gender Equal Society Perspective: Lessons from the Great East Japan Earthquake*, Tokyo: Cabinet Office, Government of Japan, June.

Goldenberg, S. 2011, 'For Japan disaster survivors, trust is as badly destroyed as the landscape', *Guardian*, 20 March, http://www.theguardian.com/world/2011/mar/20/japan-disaster-survivors-lose-trust, viewed Dec. 2013.

Hamada, K. 2013, 'Japanese government to bear more Fukushima cleanup costs for Tepco', *Reuters*, 20 Dec., http://www.reuters.com/article/2013/12/20/japan-tepco-idUKL3N0JY1R620131220, viewed Dec. 2013.

Hasegawa, R. 2013, *Disaster Evacuation from Japan's 2011 Tsunami Disaster and the Fukushima Nuclear Accident*, Studies No. 5/13, Paris: IDDRI.

Hatamura Report 2012 (Investigation Committee on the Accident at the Fukushima Nuclear Power Stations of TEPCO), http://icanps.go.jp/eng, viewed Dec. 2013.

Hobson, C. 2014a, 'Human security after the shock: Vulnerability and empowerment', in C. Hobson, P. Bacon and R. Cameron (eds), *Human Security and Natural Disasters*, London: Routledge, pp. 22–36.

Hobson, C. 2014b, 'Human security and *fortuna*: Preparing for natural disasters', in C. Hobson, P. Bacon and R. Cameron (eds), *Human Security and Natural Disasters*, London: Routledge, pp. 181–8.

Hobson, C., Bacon, P. and Cameron, R., eds. 2014, *Human Security and Natural Disasters*, London: Routledge.

ICESCR Gender Report Project 2013, *Japan: Briefing to the UN Committee on Economic, Social and Cultural Rights: Gender Report*, New York: UN Committee on Economic, Social and Cultural Rights, 50th Session, April.

Japan Ministry of Health, Labour and Welfare 2013, 'Information on the Great East Japan Earthquake: food', http://www.mhlw.go.jp/english/topics/2011eq/index_food.html, viewed Dec. 2013.

Japan Reconstruction Agency 2013, 'Current status and path toward reconstruction', May, http://www.reconstruction.go.jp/english/130528_CurrentStatus_PathToward_FINAL.pdf, viewed Dec. 2013.

Kantei n.d., '3.11 tsunami debris', http://www.kantei.go.jp/jp/singi/kaiyou/hyouryuu/qanda_eng.html, viewed Dec. 2013.

Maeda, R. 2012, 'Japan video shows delay in using seawater to cool meltdown reactor', *Reuters*, 14 Aug., http://www.reuters.com/article/2012/08/14/us-japan-nuclear-idINBRE87D0F720120814, viewed Dec. 2013.

Mainichi 2013a, 'Contaminated waste at public apartments shocks Fukushima residents', 16 Dec., http://mainichi.jp/english/english/newsselect/news/20131216p2a00m0na015000c.html, viewed Dec. 2013.

Mainichi 2013b, 'Fukushima nuclear evacuation-related deaths surpass prefecture's quake, tsunami toll', 17 Dec., http://mainichi.jp/english/english/newsselect/news/20131217p2a00m0na010000c.html, viewed Dec. 2013.

Mainichi 2013c, 'Alcoholism on the rise in tsunami-hit Miyagi Prefecture', 23 Dec., http://mainichi.jp/english/english/newsselect/news/20131223p2a00m0na006000c.html, viewed Dec. 2013.

NAIIC 2012, *The Official Report of the Nuclear Accident Independent Investigation Commission*, Tokyo: National Diet of Japan, http://www.nirs.org/fukushima/naiic_report.pdf, viewed Dec. 2013

Otsuki, N. 2012, 'Government probe: Many Fukushima hospital patients died during botched rescue operation', *Asahi Shimbun*, 24 July, http://ajw.asahi.com/article/0311disaster/life_and_death/AJ201207240092, viewed Dec. 2013.

Parker, B. 2012, 'Why did the 2011 Japan tsunami kill so many people?', *Huffington Post*, 11 March, http://www.huffingtonpost.com/bruce-parker/japan-tsunami_b_1335737.html, viewed Dec. 2013.

Plett, D. 2012, 'Swamped with saltwater: What a tsunami does to rice farmers', *The Conversation*, 12 March, http://theconversation.com/swamped-with-saltwater-what-a-tsunami-does-to-rice-farmers-5754, viewed Dec. 2013.

Polaris Project 2011, 'Polaris Japan prevents sexual violence at evacuation centers', 29 March, http://www.blog.polarisproject.org/2011/03/29/polaris-japan-prevents-sexual-violence-at-evacuation-centers, viewed Dec. 2013.

Reporters Without Borders 2013, *World Press Freedom Index 2013*, http://fr.rsf.org/IMG/pdf/classement_2013_gb-bd.pdf, viewed Dec. 2013.

Rosenberg, J. 2005, 'Globalization theory: A post mortem', *International Politics*, 42(1), pp. 2–74.

Saito, M., Takenaka, K. and Topham, J. 2013, 'Japan's "long war" to shut down Fukushima' *Reuters*, 8 March, http://www.reuters.com/article/2013/03/08/us-japan-fukushima-idUSBRE92417Y20130308, viewed Dec. 2013.

Steger, B. 2012, '"We were all in this together…" Challenges to and practices of cleanliness in tsunami evacuation shelters in Yamada, Iwate Prefecture, 2011,' *Asia-Pacific Journal*, 10/38(3), 17 Sept., http://www.japanfocus.org/-Brigitte-Steger/3833, viewed Dec. 2013.

Tabuchi, H. 2012, 'A confused nuclear cleanup', *New York Times*, 10 Feb., http://www.nytimes.com/2012/02/11/business/global/after-fukushima-disaster-a-confused-effort-at-cleanup.html, viewed Dec. 2013.

Takada, A. 2011, 'Japan's food-chain threat multiplies as Fukushima radiation spreads', *Bloomberg News*, 25 July, http://www.bloomberg.com/news/2011-07-24/threat-to-japanese-food-chain-multiplies-as-cesium-contamination-spreads.html, viewed Dec. 2013.

UNDP 1994, *Human Development Report 1994*, New York: Oxford University Press.

UNISDR 2007, *Hyogo Framework for Action 2005–2015: Building the Resilience of Nations and Communities to Disasters*, Geneva: UNISDR.

UNSG 2010, *Human Security, Report of the Secretary-General*, UNSG A/64/701, 8 March, http://hdr.undp.org/en/media/SG_Human_Security_Report_12_04_10.pdf, viewed Aug. 2013.

UNSG 2012, *Follow-up to General Assembly Resolution 64/291 on Human Security: Report of the Secretary-General*, UNSG A/66/763, 5 April, https://docs.unocha.org/sites/dms/HSU/Publications%20and%20Products/Reports%20of%20the%20Secretary%20General/A-66-763%20English.pdf, viewed July 2013.

Watts, J. 2011, 'Japan tsunami: A shortage of petrol and food but many bodies to bury', *Guardian*, 15 March, http://www.theguardian.com/world/2011/mar/15/japan-tsunami-petrol-food-bodies, viewed Dec. 2013.

Yamashita, A. 2013, 'On women in disaster-affected areas and human rights, sexual orientation and gender identity', NGO statement on behalf of Rise Together and Gay Japan News, 50th Session of the UN Committee on Economic, Social and Cultural Rights, 29 April.

2 The politics of human security in Japan

Paul Bacon

Introduction

The Japanese government has been one of the leading proponents of human security since its introduction in 1994. It has brokered and sponsored the UN Human Security Trust Fund and the Commission on Human Security, and has developed a conception of human security which it has consistently sought to apply to developing countries. The Japanese government has attempted to mobilize the international community in support of this conception, and has also explicitly sought to diffuse its expertise in disaster management. According to the 2010 Japan International Cooperation Agency (JICA) annual review,

> Japan is an advanced country in the field of disaster prevention and has compiled numerous achievements in this area. ... Japan is the only country capable of integrated comprehensive cooperation to provide continuous seamless support ranging from disaster emergency response to recovery and reconstruction as well as building disaster prevention structures. JICA regards its assistance for disaster response and prevention as a type of intellectual property that Japan should transmit to the world, and will strengthen its efforts in this area as a form of support that benefits the international public interest within the global community.
>
> (JICA 2010, p. 18)

On this view, disaster mitigation expertise is something to be transmitted outward to the world, in the international public interest, to the global community. The JICA Issue-Specific Guidelines for Disaster Reduction explicitly identify the fact that disaster prevention is considered to be a human security issue by the Japanese government (2007, p. 14). We might therefore infer that Japan's response to its own triple disaster can and should be considered a human security issue. And yet this has not been the case. It is clear, and deeply unfortunate, that Japan was unable or unwilling to bring the disaster mitigation expertise accumulated within JICA to bear on its own triple disaster. According to Yukie Osa, director of a leading Japanese NGO, the Japanese government

> was not accustomed to being a recipient and was not in a position to utilize the UN system for disaster relief, which Japan has been supporting for a long time. Also, the expertise and knowledge in overseas disaster relief built up over the years by the Japanese Ministry of Foreign Affairs (MOFA) as well as JICA, which is in charge of official development assistance, were hardly used at all following 3/11, since the Cabinet Office was in charge of relief activities. MOFA was limited to acting as a liaison for the influx of foreign aid.
>
> (Osa 2012, p. 12)

Why was Japan's hard-won expertise not utilized internally, to address 3/11? Much of the answer lies in the way that Japan has interpreted and politicized the concept of human security, and has used it to focus on overseas development assistance (ODA), in the broader sense of that term. Human security has emerged as a concept that serves to differentiate between inside and outside, and signifies what Japan is doing or ought to be doing externally, with the result that human security is not something considered to be conceptually relevant and therefore practically applicable within Japan. This framing of human security has prevented Japan's expertise from being brought to bear on its own natural disaster. Avoidable harm has occurred within Japan as a result. This is deeply unfortunate at the human level, and inconsistent with the notion, established in the 1994 Human Development Report (HDR), that human security is preventive, people-centred, universal and interdependent.

This chapter seeks to explain why the Japanese government took up human security, how the 'Japanese' conception of human security developed and prevailed, and what the costs have been. In the context of the book as a whole, the chapter performs a ground-clearing exercise, illustrating how Japan came to favour a development-based conception that it only, mistakenly, considered relevant to the world beyond Japan. Subsequent chapters perform the more positive task of demonstrating how the more comprehensive and inclusive concept of human security developed in the introduction to this book can successfully be applied to Japan's 'triple disasters'.

It has been suggested that the term human security should only apply when humans deliberately commit organized violence against each other (MacFarlane and Khong 2006). Against this, the 1994 Human Development Report recognizes that human insecurity can result from forces of nature such as tsunamis, as well as human action, and that human insecurity can be caused by inadequate policy choices or inaction, as well as deliberate acts of violence. The impact of forces of nature such as tsunamis can often be compounded by, and related to human action or inaction (UNDP 1994, p. 23). As noted in the *Hyogo Framework for Action*, 'disaster risk arises when hazards *interact* with physical, social, economic and environmental vulnerabilities' (UNISDR 2005, p. 1, my emphasis). Put another way, disasters occur when natural hazards and human vulnerabilities interact. Natural disasters are important catalysts, but much can depend on the quality of human action, and on human anticipation of and response to natural disasters; whether the seawalls are high enough, whether the use of nuclear energy is defensible, whether the nuclear reactors are built in the right place and whether people are taught and can manage to implement their emergency evacuation drills. If we mistakenly limit our focus solely to violence, we pay insufficient attention to the innovation, foresight and planning that can mitigate natural disasters, and to the incompetence, oversight and lack of preparedness which can exacerbate natural disasters. According to Kiyoshi Kurokawa, Chairman of the Fukushima Nuclear Accident Independent Investigation Commission, the Fukushima nuclear accident 'was a profoundly man-made disaster … that could and should have been foreseen and prevented … and its effects could have been mitigated by a more effective human response' (National Diet of Japan 2012, p. 9).

In this chapter the damaging human activity that I want to address is the Japanese political and conceptual framing of human security, and the negative outcomes that have ensued. The way in which Japan has framed human security is inconsistent with the main themes of the 1994 HDR. This has resulted in Japan's international disaster management expertise being unavailable to help Japanese disaster victims, in enduring perceptions of Japanese chequebook diplomacy at the UN and the inefficiency of the concept of human security as a useful way of adding value to the international development agenda (OIOS 2010). To explain how Japan came to frame human security in the limiting way that it has, it is necessary in the following two sections to introduce and review influential perspectives on human security, and explain when, how and why the Japanese government picked up on the concept.

'Japanese', 'Canadian' and critical perspectives on human security

It has been argued by many commentators that there are two main perspectives on human security: 'Japanese' and 'Canadian'. The Japanese perspective has come to be closely associated with the economic, social and cultural rights agenda, and draws on the broad and inclusive definitions of human security offered in both the 1994 HDR, and the 2003 report of the Japanese-sponsored Commission on Human Security, *Human Security Now*. In the 1994 report, the UNDP broadly defined human security as 'freedom from fear and freedom from want'. Human security is identified there as having four basic characteristics; it is universal, it is people-centred, threats to human security are interdependent, and there is a need to focus on early prevention (1994, pp. 22–3). It is suggested that human security is more easily identified through its absence than its presence, and that it has two characteristics: safety from chronic threats such as hunger, disease and repression, and, secondly, protection from sudden and hurtful disruptions in the patterns of daily life. There are seven key components of human security: economic security (freedom from poverty); food security (access to food); health security (access to health care and protection from diseases); environmental security (protection from such dangers as environmental depletion and pollution); personal security (physical safety from things such as torture, war, criminal attacks, domestic violence, drug use, suicide and traffic accidents); community security (survival of traditional cultures and ethnic groups as well as the physical security of these groups); and political security (enjoyment of civil and political rights, and freedom from oppression) (1994, pp. 24–33).

The report identifies environmental hazards as a significant human security problem (UNDP 1994, p. 22), and natural disasters are discussed at some length during the discussion of environmental security, where it is noted that poor people are much more exposed to disasters than rich people, that poor nations are more vulnerable to natural disasters and that these disasters are an integral part of the poverty cycle in developing countries (p. 29). The report concludes that 'only sustainable human development – which increases the security of human beings and of the planet we inhabit – can reduce the frequency and impact of natural disasters' (1994, p. 29). We can see here an early awareness of uneven vulnerability and of the need for sustainable responses, and an explicit connection of human security to the development agenda. Despite the headline commitment to the notion that human security is universal, we can also perhaps discern an early implication in the report that human security is considered to be more relevant to developing than developed countries.

The Commission on Human Security (CHS) report, the background to which I will discuss in more detail below, defined human security as protecting:

> the vital core of all human lives in ways that enhance human freedoms and human fulfillment. Human security means protecting fundamental freedoms – freedoms that are the essence of life. It means protecting people from critical (severe) and pervasive (widespread) threats and situations. It means creating political, social environmental, economic, military and cultural systems that together give people the building blocks of survival, livelihood and dignity.
>
> (2003, p. 4)

MacFarlane and Khong, in their highly influential text, characterize the report by the Commission on Human Security as 'the definitive statement of the development approach to human security' (2006, p. 159), and suggest that, although the report tries to strike a balance between the development and protection dimensions, it actually places 'a strong emphasis on economic welfare and its safeguarding as a fundamental component of human security' (p. 160). MacFarlane and Khong conclude that 'one sees in the CHS report an almost explicit rejection of the narrower focus on protection from violence and physical threat, and an insistence that it made little sense, when discussing security, to ignore broader social, economic and environmental threats to human well-being' (pp. 160–1). From this 'Japanese' perspective, it should be noted, the use of force is viewed with the utmost suspicion.

The Canadian perspective on human security is closely but not exclusively associated with the notion of the responsibility to protect (R2P), first set out in the 2001 report of the International Commission on Intervention and State Sovereignty, which was conceived and funded by the Canadian government.

The focus is on the individual rather than the state, and three responsibilities are identified: the responsibilities to prevent, react and rebuild, respectively. There is an emphasis on conflict prevention, and a preparedness to endorse the use of force as a legitimate policy instrument where a responsibility to react is thought to exist. The idea of human security is viewed, on this definition, as applying to a smaller, more clearly defined and specific set of political cases, with a high threshold before the use of force is appropriate. These cases are genocide, war crimes, ethnic cleansing and crimes against humanity. MacFarlane and Khong broadly endorse this Canadian perspective, identifying human insecurity as arising from organized human attempts to do harm to other humans (2006, p. 245). However, we should note that the original concept of a responsibility to protect has been considerably narrowed and sharpened, in official documents such as the 2005 World Summit Outcome Document. This sharper, more exclusive definition of the R2P has been referred to in Security Council Resolutions, and was specifically invoked as a justification for intervention by the international community in the recent case of Libya (Bellamy and Williams 2011, p. 825). It is also important to note that the current Canadian government has a very different character to that of the earlier Canadian government that sponsored the original report and commission.

There is also a set of recurring criticisms of the concept of human security which its proponents must address. According to its critics, human security lacks a precise, agreed upon definition that would allow academics to identify what to study and what to create research questions about, and which would allow practitioners to identify and prioritize some goals over others (Paris 2001, p. 92). Many argue that the landmark human security definitions cited above are somewhat vague. The Commission on Human Security refrains from proposing an itemized list of what makes up human security, because 'what people consider to be "vital" – what they consider to be "of the essence of life" and "crucially important" – varies across individuals and societies' (CHS 2003, p. 4). Paris also argues that the concept is useful only as a unifying label for those who prioritize development and humanitarian issues, that the vagueness in the definition of human security is deliberate. Paris goes so far as to suggest that the ambiguity of definition is cultivated in order that the term human security can hold together a fragile coalition of middle powers, international agencies and NGOs, and be an effective campaign slogan with which to attract public attention, political interest and government money (2001, p. 95). According to Johann Cels, one of the drafters of the 2003 CHS report, human security was originally intended to be a policy-relevant concept, rather than a theory (Cels 2012). It can therefore be argued that human security is a useful general slogan for mobilizing resources, and a useful general concept, rather than something which is capable of being developed into a theory.

There are four further criticisms which relate to the mainstreaming and implementation of human security. First, many governments, IGOs, NGOs and other agencies could be said to be implementing a human security agenda, but they are not explicitly calling what they 'do' human security, because they feel that they have already been doing it for decades (see e.g. Edstrom 2011, pp. 19–20, discussed below). Similarly, many governments, IOs, NGOs and other agencies could be said to be implementing a human security agenda, and do refer briefly to the concept, but do not mainstream or emphasize it particularly. A good example of this is the JICA Issue-Specific Guidelines for Disaster Reduction (2007), which refer to human security several times in the introduction, but make no systematic effort to integrate the concept into the technical discussion in the guidelines proper. Thirdly, because there is no clear definition of human security, it is difficult to provide meaningful bureaucratic benchmarks, and evaluate how successfully the concept is being implemented. Finally, there is the risk of institutional duplication and rivalry in the implementation of the human security agenda at the UN (OIOS 2010).

The Japanese government picks up human security

Although one of the main human security approaches has come to be labelled the 'Japanese' perspective, it can be argued that the extent to which the Japanese government has considered human security a central pillar or key element of its foreign policy has been systematically exaggerated over the years. Although it has become a truism to suggest that human security has become a pillar or key feature of Japan's foreign policy, Edstrom (2011) argues that human security ceased to be a core element of Japanese foreign policy as early as 2003.

Japan did at one point enthusiastically associate itself with the concept. Former Japanese Prime Minister Murayama was the pioneer in this regard, but it is former Prime Minister Obuchi who has been primarily linked to Japan's pursuit of human security. Obuchi gave a number of high-profile speeches in which he showcased and promoted the idea of human security, linking it to the ways in which Japan should address the landmines issue, and recovery from the Asian financial crisis. Obuchi also donated 500 million yen to start a Trust Fund for Human Security within the UN (hereafter UNTFHS). During the period when Obuchi was in power, it can be argued that Japan accorded high priority to human security and, even more ambitiously, that human security turned out to be the main pillar of Obuchi's foreign policy (Edstrom 2011, p. 41). When Yoshiro Mori became Prime Minister, he continued the human security policy conceived by Obuchi, making it a key concern of Japan's foreign policy. Concretely, he increased Japan's contributions to the UNTFHS to the extent that it became the largest UN trust fund, and also initiated the Commission on Human Security.

There was a significant change in the fate of human security following the accession to power of Junichiro Koizumi. In August 2003 Japan's ODA policy was reformed, and the human security concept was transferred to the ODA domain, no longer a key concern for foreign policy, as Obuchi and Mori had conceived it. In the revised ODA policy, five key concepts were enunciated, and while human security was declared to be one of the pillars of Japan's ODA policy, it was not the pre-eminent consideration. Edstrom (2011, p. 46) argues that Japan's engagement with the concept of human security has passed through four phases, which can be summarized as follows:

- Phase 1: 1999–2000 A pillar-to-be of Japan's long-term foreign policy
- Phase 2: 2001–2003 A priority of Japan's foreign policy
- Phase 3: 2004–2006 A pillar of Japan's ODA policy
- Phase 4: 2007– One of five key concepts of Japan's ODA policy

Edstrom also makes a number of other provocative arguments about the priority of human security in Japanese security discourse and practice. He argues that, even though there are a number of important commentators on human security in Japan who make a substantial contribution to the international epistemic community, there is, as such, no Japanese epistemic community on human security. He also argues that NGOs have historically had a difficult relationship with MOFA, and have only partially engaged with the Japan Platform and JANIC. What is more, some influential NGOs and members of the NGO community believe that the concept of human security is too vague to be of any practical or operational use (Edstrom 2011, pp. 19–20). The situation has improved since 2008, and MOFA has shown an increased willingness to engage with NGO groups, in an exchange of MOFA finances for NGO expertise. As a result, membership of the Japan Platform has increased, as have the financial resources available to it. However, the irony is that, while relationships between NGOs and the Japanese government have improved, overall the interest of NGOs in working under the human security banner is largely missing. To cap it all, Edstrom quotes Sadako Ogata as claiming, in an interview, that human security has never been a priority for MOFA (Edstrom 2011, p. 49).

The Japanese perspective prevails at the UN

Despite the above criticism, the broader 'Japanese' perspective is clearly reflected in two recent reports of the Secretary-General on Human Security (2010, 2012), both of which make an explicit distinction between human security and R2P, and specifically rule out the use of force. The 'Japanese' perspective has therefore emerged as the 'officially approved' version of human security, although the broadness of the Japanese concept has sometimes proved to be a vice. However, as Paris noted earlier in a more critical fashion, the broadness and lack of precision of the Japanese perspective is one of the reasons for its success in becoming the approved perspective. It is therefore important to briefly explain some of the reasons why the Japanese definition of human security was broad. Prime Minister Mori proposed that an international commission on human security be established under the UN in his speech to the UN Millennium Summit. Kofi Annan visited Tokyo in January 2001, and finalized an agreement with Sadoko Ogata to establish the Commission on Human Security, pledging that the UN would cooperate closely with the initiative. The press release following the meeting between Annan and Ogata, reprinted in the Annex of the 2003 report, clarified that the goals of the CHS were:

1 To promote public understanding, engagement and support of human security and its underlying imperatives.

2 To develop the concept of human security as an operational tool for policy formulation and implementation.
3 To propose a concrete action program to address critical and pervasive threats to human security (CHS 2003, p. 153).

It might be expected that the first task of a commission on human security would be to come up with a working definition of the concept. However, as we can see, the CHS was tasked to come up with an operational definition, and not a definition of the key concept itself. Further, the definitional tasks, such as they were, did not come first. The priority was to promote public understanding of, engagement with and support for human security. The CHS was intended to function as a kind of sales or PR agency for the concept (Edstrom 2011, p. 28). The CHS has been criticized for being unable to come up with a definition that was easy to grasp and that could be universally shared, but that, strictly speaking, is not what the CHS was tasked to do. The CHS provided not a definition, but a clarification of the objective of human security. Also, although the CHS chose the path of embracing all freedoms, the report has been characterized as placing a greater emphasis on freedom from want, with less attention given to the thornier political and military issues associated with freedom from fear, such as civil and political rights, and R2P. This made it much easier to mobilize support within Asia and in UN circles more generally for what was largely a pro-development agenda (MacFarlane and Khong 2006, p. 159).

It should be said that this broad approach to human security achieved a degree of success, in that in the early 2000s there was significant support for the concept within the Japanese government and the international community. If Japan had placed more emphasis on the R2P and had been prepared to consider relating the concept of human security to the possible use of force, this would have risked alienating important constituencies within Asia and at the United Nations. At the regional level, human security is a relatively safe policy choice and unlikely to provoke criticism from Japan's neighbours. In contrast to the human rights discourse, which is often perceived as reflecting a Western individualist ethos, human security seems far less controversial and less intrusive for Asian governments. Thus, human security in Southeast Asia is emphatically considered to be a concept that promotes freedom from want, rather than freedom from fear. As Acharya revealingly argues, the Japanese formulation of human security is 'less controversial for Asia-Pacific governments suspicious of, and uncomfortable with, the close association between human security and human rights promotion and humanitarian intervention' (Acharya 2001, p. 448).

The definition also ended up being broader because Ogata wanted the conceptual scope of human security broadened, and wanted it to have a global, rather than an Asian focus – at one point it had been suggested that the UNTFHS could be placed within the United Nations University in Tokyo (Edstrom 2011, p. 32). So, the definition remained broad partly because of the remit of the commission, partly because this approach had been successful in introducing and establishing the concept, partly because of Ogata's personal preferences and partly because Japan needed a broader definition to reassure people within the United Nations community that development would be promoted, and that the use of force would not be countenanced. The Japanese government originally intended for human security to be a referent for its ODA diplomacy in Asia. It can be argued that there was a tension between the broader ambitions that Ogata had for the CHS report, and the narrower ODA-related objectives of the Japanese government, which also placed less emphasis on human rights.

The fate of the concept of human security is also related to circumstances in the wider international political environment. Japan was under pressure in the 1990s to play a more proactive international leadership role. Human security was first suggested in the late 1990s by Japan as a way to meet this demand that it develop a new identity that went beyond bilateralism, and take on a new, more active leadership role. More cynically, the promotion of the concept of human security has been dismissed and the creation of the trust fund criticized as a new variant of chequebook diplomacy, designed to facilitate Japan's push for a Security Council seat through the middle of the 2000s (Soeya 2005, p. 211; Yokota, interviewed by Edstrom, 2011, p. 51). Critics who take this line note that Japan's support for the concept decreased commensurately with its growing realization that it would fail to secure a permanent seat on the UN Security Council. The War on Terror after 9/11 also meant that there was less emphasis placed on human security by the Koizumi administration.

To briefly recapitulate the argument: one influential way of thinking about human security has been to make a distinction between Japanese and Canadian approaches, between the development dimension and the protection dimension of human security. From a Japanese perspective, partly to reflect Japanese preferences for ODA diplomacy and partly to gain acceptance for the concept of human security within the region and at the UN, it has been necessary to adopt a general definition and approach which is pro-development, and which distances itself from questions regarding civil and political rights and humanitarian intervention involving the use of force. The Japanese perspective meets these criteria, and has therefore been accepted as the approved version of human security in UN circles. However, partly to address these political imperatives, the term has been vaguely defined and there has been a perception that the concept has been poorly operationalized at the UN level, with the result that the UNTFHS has not been a success.

Japanese and Canadian perspectives and the dangers of dichotomization

Having explained the evolution of human security debates in terms of the Japanese and Canadian perspectives, in this section I would like to briefly illustrate a number of ways in which this dichotomization has been counterproductive. The first danger is that basing discussion on the notion of the distinctiveness and incompatibility of the two perspectives is both intellectually unproductive and empirically flawed, and leads to sterile and circular debate. The second danger is that differences in emphasis between the two approaches will be exaggerated or magnified, leading to a third possibility, that of serious mischaracterization of important positions within the broader debate. This is what has happened as a result of the Japanese government treating human security as a synonym for economic development, rather than as a conception that should address civil and political rights. Political concerns such as human rights are downplayed even though they are mentioned in both reports, and the reports are also misleadingly and/or selectively interpreted as providing a pretext for the prioritization of economics over politics. A fourth problem is the possibility that important issues will be omitted from debate if perspectives are characterized in an exclusivist or exaggerated way. A good example of this is that, if the Japanese government treats human security as a synonym for the development agenda, human security will not be considered to be relevant to Japan itself or to other developed societies. Lastly, the fifth danger of dichotomization is that if the definition of human security is too broad and lacking in precision, partly as a result of political considerations, then we cannot produce practical agendas for human security promotion, which also leads to operational problems.

Framing the debate in terms of the idea of Japanese and Canadian approaches has led to the emergence of an economistic development assistance-driven conception of human security, which has precluded the possibility that the conception can be thought of by Japanese as applying domestically. Dichotomization has led to simplification and exclusion and, most importantly for our purposes, what has ended up getting excluded is the notion that human security is universal, and that it therefore also applies within Japan.

MacFarlane and Khong argue that Japan embraced sustainable human development from the early 1990s, as could be seen by its commitment to the Tokyo International Conference on African Development (TICAD) process from 1993 onwards. They also suggest that the activities supported through the UNTFHS also demonstrate a clear Japanese emphasis on human development as the central thrust of human security. Japan has supported programmes intended to alleviate poverty, provide food security, enhance the welfare of returning refugees, rehabilitate schools and hospitals, provide vocational training to ex-combatants, control disease, promote reproductive health, enhance public health services and control trafficking in women and children (2006, p. 158).

MacFarlane and Khong suggest that all of this amounts to a primarily developmentalist understanding of human security on the part of the Japanese government, and a discomfort with the interventionist thrust of the R2P concept. As noted earlier, they characterize the CHS as:

> the definitive statement of the development approach to human security … with a strong emphasis on economic welfare and its safeguarding as a fundamental component of human security.
>
> (pp. 159–61)

Furthermore, although they acknowledge that Sen's argument clearly links development to civil and political rights, MacFarlane and Khong maintain that Sen's

> vision of security (i.e., what threats people are to be protected from) is strongly economic in character … the threats identified are starvation, epidemics, and sudden and severe deprivation.
>
> (MacFarlane and Khong 2006, p. 158)

However, these are not fair readings of either the CHS report or Sen's work. The CHS report refers to a far broader human security agenda than merely development. Three of the six thematic chapters in the CHS report address people caught up in violent conflict, people on the move and people recovering from violent conflict. The CHS report actually strikes a fair balance between political and development assistance-related perspectives, and in this sense offers a far more comprehensive conception of human security than is suggested by MacFarlane and Khong. The 2003 CHS report did invoke the idea of cultural sensitivity and suggests that it was not possible to propose an itemized list of what makes up human security. However, this is significantly at odds with other parts of the report, where it is suggested that 'respecting human rights is at the core of protecting human security' (CHS 2003, p. 10). The report also strongly invokes the Universal Declaration of Human Rights, the universality and interdependence of the human rights of all people, and stresses the significance of civil and political rights (CHS 2003, p. 10). In the section which discusses empowerment, the commitment to civil and political rights is even more emphatic – it stresses that democracy and freedom of the press, belief, conscience, information and organization are all required. So, according to the CHS report taken as a whole, we need to balance our commitment to people-centredness and sensitivity to cultural difference with a strong and thoroughgoing commitment to the concept of universal human rights.

The 2003 CHS report was intended to integrate rather than separate 'freedom from want' and 'freedom from fear'. Sen famously argued that political participation is a key constitutive attribute of human development, and also that there is a clear instrumental connection between the presence of democracy and the absence of serious famine (see e.g. Sen 1999, p. 188). In both of these senses he had a very strong belief in the connectedness of civil and political rights, and economic, social and cultural rights, and that the solution is to think of such problems as requiring comprehensive conceptual integration rather than separation.

MacFarlane and Khong's characterizations of the CHS report and Sen's work have been influential, but also demonstrate what happens when we conceptualize human security in terms of Canadian and Japanese approaches. The result is a simplistic and unhelpful reductionism, and a politicized differentiation that creates exaggerated, false representations, doing little justice to important contributions that seek to demonstrate how human security concerns are complexly connected, or interdependent, as the 1994 HDR puts it.

As suggested above, there may be a difference between the balanced and comprehensive content of the CHS report, and the economic priorities of the Japanese government. This line of argument can be supported by a consideration of the impact of Ogata on the content of the CHS report. Ogata wanted to focus on broader issues than disease, poverty and underdevelopment, and had a decisive impact on the adoption of the broad approach taken by the CHS. When she was asked to co-chair the commission, she insisted on a broadening of the scope of the commission's work. In a roundtable talk, Ogata later revealed that she

> thought that the proposed focus of its work, public health, was outside her field of expertise: 'Of course disease, poverty and unemployment are among the threats to human security, but I had long been concerned with the plight of the ordinary people who were the casualties of conflict. So I made it clear that I was willing to be involved if the commission were expanded to include such issues.'
>
> (Ogata, quoted in Edstrom 2011, p. 30)

In fact, Ogata's role in widening the scope of Japan's pursuit of human security was even more significant than had been disclosed before. When Obuchi and his team were considering establishing what was to become the TFHS, their focus was Asia, as was revealed by Obuchi's speech in Hanoi in December 1998. This was in line with traditional Japanese foreign policy, with its emphasis on neighbouring Asia. But Ogata declared: 'I was not interested in an Asian focus. I saw [the pursuit] as global' (quoted in Edstrom 2011, p. 31) and the fund was established under the aegis of the United Nations in New York, which expanded its geographical scope from an Asian to a global one. So in Ogata's case we can see part of the reason why the concept was broadened, but we can also see that she was interested in integrating 'Japanese' concerns with development with 'Canadian' concerns with internally displaced people (IDPs) and ordinary victims of conflict and in creating a more comprehensive conception of human security.

Although both Ogata and Sen in their different ways tried to integrate political and economic concerns, in practice the Japanese government made the same move as MacFarlane and Khong, and chose to interpret the CHS report as giving it a mandate to focus more or less exclusively on a development assistance-related agenda. However, there is an important danger of omission associated with thinking of human security as a synonym for the development agenda. This way of thinking encourages us to think of the human security, vulnerability, resilience and empowerment discourses as relevant only to people in less developed countries. If the Japanese government associates human security with development assistance, and thinks of human security primarily as a vehicle through which to pursue a regional development strategy, it follows that the Japanese government is less likely to consider that the concept could apply to human insecurity within Japan and other developed countries.

The success of the Japanese approach?

Has the resulting Japanese approach to human security been unsuccessful? Japanese governmental practice has produced both human security and human insecurity. While the Japanese government has aimed at improving human conditions and security in Afghanistan, for example, it has concurrently produced human insecurity for Afghan refugees who tried to start new lives in Japan (Hynek 2012a, p. 132). Therefore, Japanese government practice contains two 'contradictory' elements: this 'reversed pairing' comprises a benign human security element directed at the reconstruction of Afghanistan, and a hostile element aimed against humanitarian and political refugees fleeing Afghanistan and desperately trying to get into Japan. This reveals a 'complete lack of liberal circulation of security between the inside and outside of Japan' (Hynek 2012a, p. 132). This is one sense in which the Japanese government's human security practices are contradictory, non-universal and non-interdependent in a way that compromises the human security of non-Japanese.

With regard to the human security of Japanese themselves, human security now fulfils a political role as a slogan for Japan's focus on external development assistance. In this sense, the term human security is part of a practice that differentiates between the inside and the outside, and the term's very performative function is to refer to the outside. Human security ends up as a tool of differentiation, deployed to signify that which Japan does externally, rather than internally. It is deeply unfortunate that, for the reasons traced above, the Japanese government has developed a proclivity for thinking of human security as a signifier for something external. This has precluded the possibility that a more 'authentic' conception of human security, truer to the spirit and the content of the 1994 HDR, can be conceived to apply to life within Japan. As a result, Japanese people who found their human security at risk through natural disaster were excluded from the remit of a more inclusive, interdependent and universal conception of human security.

Hynek argues that the domestic and international elements of Japan's 'reverse-paired' bureaucratic production are complementary and deliberately and systematically linked, and that Ogata has been at the heart of this plan which has enabled Japan to successfully install its human security preferences at the UN, and to use the UNTFHS to supplement its bilateral foreign policy agenda (Hynek 2012b, p. 68). The suggestion is that the Japanese government has deliberately neglected or simply failed to understand that the concept of human security could apply inside Japan, either to Japanese or to non-Japanese seeking asylum. This notwithstanding, for Hynek the Japanese conception of human security has been successful promulgated internationally, allowing Japan to dictate terms within UN development circles, and to pursue its foreign policy objectives through the UNTFHS.

However, it is difficult to imagine Japanese bureaucracies cooperating internally in such a systematic way to produce such a comprehensive and multi-faceted human security nexus. This significantly overestimates Japan's capacity for internal coordination and the often strong rivalries that exist between Japanese ministries. It is not plausible to suggest that there is a systematic human security plan which has been successfully masterminded by Ogata. It is more likely that Ogata's attempts to broaden the scope of the term to include political issues, and to broaden the geographic application of the term, were effectively sidelined by the Japanese government and MOFA, as she herself suggested in her interview with Edstrom, cited above.

Perhaps the best concrete way to assess the claim that there has been a successful and comprehensive Japanese human security strategy at the UN is to look at what UN institutions are saying, and evaluate a recent internal UN review of the UNTFHS. The UN Trust Fund for Human Security was subjected to a strongly unfavourable review in a report published in January 2010 by the UN's Office of Internal Oversight Services (OIOS). The report identified a number of problems with the fund; the money allocated to it has been underutilized; only around 60 per cent of the allocated funds have been spent; and despite the fact that the fund is a UN fund, the report estimates that almost 100 per cent of the money has come from the Japanese government (OIOS 2010, p. 3). The Japanese government has therefore operated a de facto veto on how the funds are allocated. There have been additional criticisms of the budget allocation process as being too complex, and taking far too long to administer. Perhaps because of these problems, and the issue of a Japanese monopoly, only three other countries had made recent contributions to the fund, and these had been negligible. The report noted there has been no concrete buy-in from other states (OIOS 2010, p. 4).

The report further notes that 'there has been no comprehensive evaluation of the fund's activities to determine its impact on the human security programme' (OIOS 2010, Executive Summary). The Office for the Coordination of Humanitarian Affairs (OCHA) indicated that it 'faced a major challenge identifying evaluators with the appropriate knowledge of human security to carry out an investigation' (OIOS 2010, p. 4). For the OIOS this has posed 'a significant risk exposure for the UN in not being able to demonstrate the effectiveness of the human security programme, including the effectiveness of its advocacy of the human security concept' (OIOS 2010, pp. 4–5). The report goes on to say that 'a number of trust funds and technical cooperation activities are being undertaken by the UN in similar thematic areas, and the rationale behind operating a separate trust fund with similar activities remains to be determined' (OIOS 2010, p. 5). It was not clear whether the agencies receiving resources from the fund had integrated the concept into their overall strategy, or were attempting to fulfil a unique requirement arising from human security activities (OIOS 2010, p. 4). This may have been because at the application stage 'there were no clear internal criteria for the project review and proposals' (OIOS 2010, p. 9) and also because, as stated, there has never been an adequate comprehensive review. This is clearly a damning report, which possibly even casts doubt on Paris's suggestion, above, that human security can be thought of as a useful unifying slogan or a rallying call. It also explicitly raises the issue of inter-agency competition at the UN, and the fact that the trust fund is sometimes seen as either duplicating or rivalling the work of other agencies.

Japan has developed its own conception of human security, and has attempted to use it as a device to achieve Japanese foreign policy and development-assistance objectives. But it has arguably failed practically to achieve these objectives, as the OIOS report suggests. Importantly for our purposes, the Japanese approach has also failed in the sense that the possibility that human security can be thought to apply within Japan has been precluded. Japan has therefore been unfaithful to the 'philosophy' of human security, as espoused in the 1994 HDR, by politicizing the phrase and using it to signify that which happens externally, thereby committing itself to a version of the concept that does not reflect the claim and demand that human security be preventive, people-centred, universal and interdependent. The overall conclusion must be that the Japanese approach has not adequately provided for the achievement of human security within or beyond Japan's borders, and that it is surely time for a more inclusive rethink of the way in which the Japanese government conceives of human security.

Conclusion: towards a human security approach to natural disasters within and beyond Japan?

The primary purpose of this chapter has been to attempt to explain why the concept of human security has not hitherto been thought to be applicable within Japan. Even though Japan prides itself on its human security expertise, especially with regard to disaster mitigation, it was not able to bring this expertise to bear on its own triple disaster. 'Natural' disasters often have a significant human element to them. The impact of natural disaster can be significantly mitigated by the right kind of human anticipation and policy, and significantly exacerbated by the wrong kind of human action. Slightly abstractly, the way that Japan has framed human security as a concept that applies only to external affairs is itself a significant act of social construction, a human error which has increased suffering in relation to the triple disaster. I have tried to emphasize in this chapter that Japanese political choices and moves have played an important role in framing the 'Japanese' perspective, shaping the dichotomized debate, and contributing directly to the resultant strengths and weaknesses of the concept of human security. Studying when, how and why the Japanese conception emerged in its current politicized and exclusionary form is in itself an important if somewhat abstract way of engaging in disaster mitigation – by identifying and eliminating the collective human error which led to the establishment of a conception that removed Japanese themselves from the legitimate purview of human security.

The chapter has therefore performed a critical and preventive theoretical ground-clearing exercise. Subsequent chapters assume the more positive task of showing how the concept of human security and the vulnerability, resilience and empowerment discourses are connected to each other, and how they can be meaningfully and helpfully applied to Japan. It is necessary to develop a positive human security agenda for Japan that emphasizes the 'human' element to human security, and acknowledges the role that humans play in mitigating or exacerbating natural hazards. This approach should be universal, as required by the 1994 HDR report, and should therefore apply equally to those within and without Japan. The interdependence of the different dimensions and elements of human security should be emphasized and spelt out. This positive agenda should move beyond or downplay the unfruitful dichotomizing frame of the Canadian/Japanese approach to human security. The critical potential of the idea of prevention, another key dimension of human security according to the 1994 HDR, should be developed in the context of debate about the desired future Japanese energy mix. The vulnerability, resilience and empowerment literatures provide a framework within which we can analyse the people-centred or human dimension of human security, and address rebuilding issues, such as the role of local festivals and rituals in helping people to restore a sense of community, the post-trauma psychological recovery of the Fukushima workers, and the logistical, ethical and gender issues related to the provision of emergency housing and resettlement.

Perhaps, more generally, this is an important opportunity for the Japanese government to reaffirm its commitment to human security, and to 'reboot' the concept by focusing more explicitly on the connection between human security and natural disasters. Instrumentally, this could be a good way of addressing some of the criticisms of the UNTFHS contained in the OIOS report, by making the concept more focused, quantifiable and prescient, and by inviting other countries to participate more actively in the financing and administration of the fund. Natural disasters are increasingly frequent and destructive phenomena, and there would of course be a significant poignancy and symbolism to Japan focusing more attention on and assigning more resources to natural disasters, both within and beyond its own borders, in the wake of 3/11. Perhaps the third anniversary of the triple disaster, and the twentieth anniversary of the 1994 HDR provide an important opportunity for Japan to show leadership in response to its own crisis, and humility and inclusiveness in recognizing its own vulnerability and the vulnerability of all to human insecurity. There may never be a more appropriate moment for Japan to commit to a more comprehensive conception that genuinely embraces a commitment to all four dimensions and all seven elements of human security, and seeks to empower all of those whose human security is threatened by natural disaster.

References

Acharya, A. 2001, 'Human security: East versus West', *International Journal*, 56(3): pp. 442–460.

Bellamy, A. and Williams, P. 2011, 'The new politics of protection? Côte d'Ivoire, Libya and the responsibility to protect', *International Affairs*, 87(4): pp. 825–850.

Cels, J. 2012, 'Natural disasters and forced population displacements: A human security perspective', Research Workshop: Human Security and Natural Disasters, Waseda University, Tokyo, 20 Feb.

Commission on Human Security. 2003, *Human Security Now*, New York: CHS, http://isp.unu.edu/research/human-security/files/chs-security-may03.pdf, viewed Dec. 2013.

Edstrom, B. 2011, *Japan and Human Security: The Derailing of a Foreign Policy Vision*, Stockholm: Institute for Security and Development Policy, Asia Papers.

Hynek, N. 2012a, 'The domopolitics of Japanese human security', *Security Dialogue*, 43(2): pp. 119–137.

Hynek, N. 2012b, 'Japan's return to the chequebook? From military peace support to human security appropriation', *International Peacekeeping*, 19(9): pp. 62–76.

International Commission on Intervention and State Sovereignty. 2001, *The Responsibility to Protect*, Ottawa: IDRC.

JICA. 2007, *Issue-Specific Guidelines for Disaster Reduction*. Tokyo: JICA.

JICA. 2010, *Boosting Growth, Aiming for Inclusive Development*, Annual Report. Tokyo: JICA.

JICA. 2011, *Assessment of JICA Contribution to the Hyogo Framework for Action*. Tokyo: JICA.

MacFarlane, N. and Khong, F. K. 2006, *Human Security and the UN: A Critical History*, Bloomington, IN: Indiana University Press.

National Diet of Japan. 2012, *The Fukushima Nuclear Accident Independent Investigation Commission*, Tokyo: National Diet of Japan.

Office of Internal Oversight Services. 2010, *Management of the United Nations Trust Fund for Human Security: The Impact of the United Nations Trust Fund for Human Security Programme Remains Undetermined*, Internal Audit Division, Audit Report, Assignment No. AN2009/590/04, 29 Jan.

Osa, Y. 2012, 'Human security vs. national security in a developed country: Japan's triple disaster and the challenges of Japanese civil society', Research Workshop: Human Security and Natural Disasters, United Nations University, Tokyo, 21 Feb.

Paris, R. 2001, 'Human security: Paradigm shift, or hot air?' *International Security*, 26(2), pp. 87–102.

Sen, A. 1999, *Development as Freedom*, Oxford: Oxford University Press.

Soeya, Y. 2005, *Nihon no 'midoru pawā' gaikō: Sengo Nihon no sentaku to kōsō* (*Japan's 'Middle Power' Diplomacy: Postwar Japan's Choices and Conceptions*), Tokyo: Chikuma shobō.

UNDP. 1994, *Human Development Report 1994*. New York: Oxford University Press.

UNGA. 2005, 2005 *World Summit Outcome*, A/60/1. New York: UN.

UNSG. 2010, *'Human Security', Report of the Secretary-General*, UNSG A/64/701, 8 March, http://hdr.undp.org/en/media/SG_Human_Security_Report_12_04_10.pdf, viewed Dec. 2013.

UNSG. 2012, *Follow-up to General Assembly Resolution 64/291 on Human Security: Report of the Secretary-General*, UNSG A/66/763, 5 April, https://docs.unocha.org/sites/dms/HSU/Publications%20and%20Products/Reports%20of%20the%20Secretary%20General/A-66-763%20English.pdf, viewed Dec. 2013.

UNISDR. 2005, *Hyogo Framework for Action 2005–2015: Building the Resilience of Nations and Communities to Disaster*, Geneva: UNISDR.

UNISDR. 2012, *Towards a Post-2015 Framework for Disaster Risk Reduction*, Geneva: UNISDR.

3 Mismanaging risk and the Fukushima nuclear crisis

Jeff Kingston

Introduction

> Though global safety standards kept on improving, we wasted our time coming up with excuses for why Japan didn't need to bother meeting them.
>
> (Haruki Madarame, Chairman, Nuclear Safety Commission, Diet testimony, 15 Feb. 2012)

The nuclear accident at Fukushima was precipitated by natural disaster, but poor risk management, including a failure to comprehend the lessons of disasters in the most earthquake-prone country in the world, along with institutionalized complacency about risk, were major factors increasing the likelihood of a major accident and fumbling crisis response. Tokyo Electric Power Company (TEPCO), the utility operating the Fukushima Dai-ichi plant, and the Nuclear and Industrial Safety Agency (NISA), the government regulatory authority, mismanaged a range of risks – siting, seismic, tsunami, emergency preparedness and radiation contamination – and it was this mismanagement that made Fukushima into Japan's Chernobyl. Investigations into the accident have established that the crisis response was improvised and inadequate because of lack of preparation, institutional flaws in emergency procedures and poor communication within the government and between officials and TEPCO. The tunnel vision of vertical administration within the fractured government bureaucracy, where turf and jurisdiction are jealously guarded, was a major factor in the institutional paralysis that slowed the crisis response during and after 3/11.

Human security, as it turns out, was vulnerable to the interplay of natural and nuclear disasters and institutionalized failures in preparing for worst-case scenarios and nurturing a culture of safety where it should have been a priority. As the tragedy unfolded, the key institutions and actors failed to communicate with each other and the emergency response was uncoordinated and ad hoc. This grave series of failures is all the more alarming because it happened in Japan, a nation known for its meticulous disaster emergency planning. The lessons of Fukushima are still being digested, but at the very least should jolt any and all governments and international organizations out of any misplaced complacency and raise awareness about multi-hazard risk.

About six weeks after the 3/11 complex disaster, the Japanese government submitted a progress report on 25 April 2011 regarding its implementation of the Hyogo Framework discussed in the introduction to this volume. The Cabinet Office report was probably drafted largely before the cascading disaster of earthquake, tsunami and nuclear reactor meltdowns, but it is curious that the events and aftermath of 3/11 are not even mentioned. More significantly, the Japanese government's self-assessment is quite positive and details various policy achievements in terms of addressing the Hyogo Framework's five priorities for action. For example, see the section on the Hyogo Framework priority for action 5: strengthen disaster preparedness for effective response at all levels. Under this section, the Cabinet reports, 'Comprehensive achievement with sustained commitment and capacities at all levels' (Cabinet Office 2011, p. 20). This positive assessment is followed by details on the legal and institutional framework that is designed 'to promote emergency measures, rehabilitation and reconstruction activities in a coordinated manner among larger areas in a large-scale disaster' (Cabinet Office 2011, p. 21). The report also suggests that disaster preparedness plans, and contingency plans are in place, reinforced by training drills and rehearsals.

By the time this Cabinet progress report was submitted it was abundantly clear that the emergency response to the Fukushima nuclear accident was grossly inadequate by any measure. There had been no evacuation drills or disaster training because TEPCO, the operator of the Fukushima nuclear plants, had not wanted to alarm nearby residents. The government had not interceded to insist on such drills because it was complicit in perpetuating the myth of 100 per cent safety that the nuclear industry had propagated to promote the expansion of nuclear power. Moreover, the government had developed a computer simulation system designed to predict radiation dispersion called SPEEDI so that it could guide evacuations of residents in vulnerable communities to areas of safety. In the event, the government did not use the SPEEDI data until late April, but only after it had evacuated residents in communities of relatively low exposure to radiation hotspots, needlessly exposing thousands of evacuees to far higher levels of radiation than would have been the case if they had remained in their homes.

Residents of Japan watched the Fukushima disaster unfold live on their televisions, complete with hydrogen explosions that sent plumes of radiation into the atmosphere, and were certainly not reassured by images of helicopters trying to drop water onto the reactors and spent fuel pools, more often missing than not. In the Japanese media there has been a relentless cascade of revelations about fundamental errors, lax enforcement of safety guidelines, shady practices and negligence that has spawned an anti-nuclear energy movement driven by citizens' newfound awareness about how the government and utilities were poorly prepared to cope with a disaster that should have been a priority concern. Japanese people understand that human security was compromised to accommodate nuclear industry interests.

The Japanese Cabinet Office's Hyogo Framework progress report never addresses just how woefully unprepared the government was for a complex, cascading catastrophe on the scale of 3/11. The government had checked off the boxes, put the legal and institutional framework in place, but had neglected to imagine and prepare for worst-case scenarios. There was an institutionalized myopia and wishful thinking about the safety of operating nuclear reactors in one of the most earthquake- and tsunami-prone places in the world. Investigations into the disaster have revealed that TEPCO and the government systematically downplayed risk and avoided expensive countermeasures due to industry lobbying. The nuclear industry lacked a culture of safety because of regulatory capture; government regulators regulated in favor of the regulated.

The problems of regulatory capture lay outside the Hyogo Framework, but are part of the seamy, widespread political realities that influence disaster preparedness and emergency responses all over the world. This is particularly true of the nuclear industry where the vested interests in government and industry seem more concerned with overcoming opposition to nuclear energy than in adopting strict safety guidelines and enforcing them (Kingston 2012c).

Three major investigations into the nuclear disaster concluded that TEPCO's systematic negligence contributed to the nuclear disaster and criticized its 'make-believe' disaster emergency arrangements. Yoichi Funabashi, former editor of the *Asahi* newspaper and prominent public intellectual, launched a non-government investigation conducted by the Rebuild Japan Initiative Foundation under the direction of Koichi Kitazawa, former chairman of the Japan Science and Technology Agency. The Hatamura investigation was a third-party panel initiated by the government that was headed by noted disaster expert and university professor Yotaro Hatamura with the support of various experts, many from government institutions. The National Diet investigation was the first independent inquiry conducted at the Diet's behest and headed by Kiyoshi Kurokawa, former president of the Science Council of Japan (Funabashi Report 2012; Hatamura Report 2012; National Diet 2012). The myth that nuclear reactors could be operated with absolute, 100 per cent safety, embraced and promoted by the nuclear village of pro-nuclear advocates, made it taboo to question safety standards and militated against sober risk assessment and robust disaster emergency preparedness. Those responsible for operating or regulating nuclear reactors propagated this myth of 100 per cent safety and were blinded by it, a collective failure that left them unprepared to deal with an accident. Paradoxically, this safety myth explains why TEPCO lacked a culture of safety and why its crisis response was so deficient. Moreover, politicians lacked knowledge about nuclear issues and crisis management, and did not get sufficient support or information from bureaucrats or TEPCO to cope with the crisis. This failure to share information bred mistrust between key actors and institutions that impaired an effective response.

This chapter examines how TEPCO minimized risk assessments and preparations prior to 3/11, how it has tried to shirk and shift blame since then, and how the government's mismanagement of the nuclear industry continues to imperil human security in Japan. Elsewhere I have documented TEPCO's efforts to blame former PM Naoto Kan for its own miscues and failure to prepare adequately for the evident risks (Kingston 2011b). As I explore below, the nuclear village of pro-nuclear advocates had much to gain by shifting blame to Kan and diverting attention from the institutional problems that are at the heart of the crisis.

Back in 1994 the UNDP defined the concept of human security, asserting that it is something that people instinctively recognize and concerns sudden and hurtful disruptions in the pattern of our daily lives, including environmental harm (UNDP 1994, p. 3). The residents of Fukushima have encountered abrupt and severe disruption of their lives as tens of thousands have been evacuated and many face the prospect of decades of displacement. Residents of the prefecture must cope with huge financial losses, loss of community and livelihood, separation of families, social stigma and the uncertainties of the health consequences of exposure to radiation. The Fukushima accident was an avoidable and egregious violation of human security throughout Japan, as people all over the archipelago suddenly understood that their trust in government and deference to the experts and authorities has been mistaken in the case of nuclear energy.

A record magnitude 9 earthquake and subsequent 15-meter tsunami pulverized the Tōhoku coastline on 11 March 2011, claiming some 20,000 lives, wiping out entire towns and inundating the Fukushima Dai-ichi nuclear plant. These extreme seismic events were the proximate causes that led to the loss of electricity and the failure of backup generators. The ensuing cessation of the cooling systems caused three meltdowns within the first 80 hours and the hydrogen explosions that released plumes of radiation, spreading radioactive contamination in surrounding areas but also further afield due to strong spring winds. The long-term health effects are uncertain, but the costs of the nuclear crisis have been enormous and are mounting. The reckoning includes displacement of some 80,000 residents within the 20 km evacuation zone around the crippled reactors, many of whom will probably never return to their homes, and massive losses suffered by local farmers, fishermen and various businesses in Fukushima due to the stigma of radiation that envelops the prefecture. This stigma follows those who leave to restart lives elsewhere and raises concerns among young people concerning marriage prospects and raising families. In addition, there has been a wider economic fallout as bans on Japanese agricultural products were imposed overseas, tourists refrain from visiting and the nuclear plants no longer generate local government revenues. Moreover, the nuclear crisis tarnished the Japan-brand, eroding the nation's reputation for technological prowess. Claims for compensation are mounting while the costs of decontamination, disposal of tainted debris and decommissioning nuclear reactors will boost the final reckoning considerably. In addition, taxpayers are funding a massive government bailout of TEPCO, totaling nearly $45 billion as of mid-2012 (*Wall Street Journal*, 5 Dec. 2012). Future generations of Japanese will be paying for the folly of Fukushima for quite some time.

It is important to learn lessons from the poor risk management in the nuclear industry because Japan will continue to rely on nuclear energy for years to come despite the Fukushima debacle. In May 2012, all of Japan's 50 viable reactors were offline for safety checks, marking the first time in four decades that Japan was not depending on nuclear power for any electricity. Controversially, the Noda Cabinet decided to restart two reactors in Oi during the summer of 2012, ostensibly to ensure stable supplies of electricity and lessen imports of fossil fuels. Public opinion polls indicated that a majority of Japanese opposed restarting nuclear reactors in mid-2012, and there are widespread anxieties about the operational safety of nuclear reactors, but Prime Minister Yoshihiko Noda overrode such objections. He also ignored several mass street protests across from his office, the largest demonstrations in central Tokyo since the 1960s. Polls show that from two-thirds to three-quarters of the public wants to eliminate or reduce nuclear energy, but the utilities have invested vast sums in this option, one made possible only by vast government subsidies, and oppose pulling the plug. Moreover, it will take considerable time and money to ramp up renewable energy generating capacity. While renewable energy may have a promising future in Japan, in the meantime it is important that regulators and operators minimize the inherent risks of operating nuclear plants in a seismically active nation by acting on the lessons of Fukushima and implementing more stringent safety measures and improved crisis response procedures. A defining component of human security is freedom from fear, which is usually understood in reference to conflict and violence, but in Japan the fears generated by 50 viable nuclear reactors, and concerns about safely disposing of nuclear waste, are widespread as many citizens fear the government and utilities are ignoring the lessons of Fukushima at their collective peril.

Embracing risk

Fukushima was preceded by a series of mishaps, cover-ups, irresponsible practices, close calls and ignored warnings. In that sense, it was an accident waiting to happen. Charles Perrow has written extensively on the inevitability of accidents in organizations predicated on complex technologies and the problem of unexpected interactions that may cause a cascading disaster as occurred at Fukushima. He writes,

> some complex organizations such as chemical plants, nuclear power plants, nuclear weapons systems … have so many nonlinear system properties that eventually the unanticipated interaction of multiple failures may create an accident that no designer could have anticipated and no operator can understand. Everything is subject to failure-designs, procedures, supplies and equipment, operators, and the environment. The government and businesses know this and design safety devices with multiple redundancies and all kinds of bells and whistles. But nonlinear, unexpected interactions of even small failures can defeat these safety systems. If the system is also tightly coupled, no intervention can prevent a cascade of failures that brings it down.
>
> (Perrow 2011)

Given this apparent inevitability of accidents, and the fact that Japan suffers 20 per cent of the world's >6 magnitude earthquakes and coined the term 'tsunami', it may seem surprising that the government decided to place such a big bet on nuclear energy and decided to construct clusters of multiple reactors that amplify the risks. Certainly the oil embargoes and price hikes of the 1970s reinforced perceptions that Japan had no choice. The nuclear fuel cycle was pursued because it offered the hope of significantly reducing Japan's dependence on energy imports. And as more money was invested in expanding Japan's network of nuclear power plants, it created vested interests in the government and utilities committed to further expansion. This 'nuclear village' of pro-nuclear advocates, comprising the utilities, bureaucracy, politicians, mass media and academia, is disinclined to reexamine underlying assumptions about whether it is possible to operate nuclear reactors safely in such a seismically active area. And as time passed and no major mishaps occurred, nuclear advocates became increasingly blasé about the risks and focused narrowly on the benefits of a reliable, relatively inexpensive energy source. Moreover, as concerns about global warming grew towards the end of the twentieth century, advocates discovered a new reason to promote expansion of nuclear energy: it contributes to the goal of reducing carbon emissions. Given the clear environmental and human costs associated with reliance on carbon fuels, ranging from extraction, processing, transporting and emissions, nuclear energy has its advantages. So over the decades, nuclear power developed an accumulating and appealing logic that relied on downplaying the problems related to disposing of radioactive waste and the potential for accidents due to human error, natural disaster or a combination thereof.

As Daniel Aldrich argues in *Site Fights* (Aldrich 2008), there is a keen appreciation in the nuclear industry that the public needs convincing about the safety of nuclear power. This explains why communities that host reactors are lavished with financial inducements. Aldrich explains that communities with low levels of social capital are specifically chosen in order to reduce the risk of local opposition and because their marginal socio-economic situation makes them more inclined to accept lucrative offers. From the perspective of human security, economic insecurity leaves them more vulnerable and with less freedom of choice. Hiroshi Onitsuka reveals that the deep pockets of the central government and the utilities bestow benefits on hosting communities prior to construction, creating a subsidy addiction (Onitsuka 2012). Jobs, tax income, various subsidies and extravagant public facilities are combined with reassuring public information campaigns to assuage concerns and build support for nuclear power projects. In Japan's declining remote coastal towns and villages, it is understandable that the risk of poverty and bleak futures outweighed the potential risk of nuclear energy. Deferential views toward the central government, together with a pragmatic assessment that such projects will be built somewhere and someone will benefit, also help explain why hosting seemed a reasonable option.

Although the central government and utilities promoted a nuclear consensus – nuclear energy is safe, reliable and cheap – some civil society groups and many individual Japanese contested this effort to little avail. There was no national anti-nuclear energy movement pre-3/11 and the anti-nuclear bomb activists did not embrace this issue (Avenell 2012). As a result, the 'nuclear village' has dominated the conversation. These advocates are not given to doubts or inclined to reconsider their assumptions and have relied on their power networks to prevail. Prior to Fukushima there were 14 lawsuits challenging nuclear power plants on the grounds that seismic dangers were hidden or downplayed, but the utilities prevailed in each case (Repeta 2011).

The utilities, government and associated scientists tout the high-tech, fail-safe features of nuclear reactors, but as Perrow reminds us, accidents happen. Immediately after the 11 March disaster, TEPCO was quick to claim that the tsunami and chain of multiple failures were inconceivable, but the record suggests otherwise. In 1975, nuclear chemist Jinzaburō Takagi and others established the Citizens Nuclear Information Center (CNIC) and since then issued regular reports on nuclear power plant safety issues. This activism targeted the regulatory and technical problems with nuclear power and the vulnerabilities specific to seismically active Japan. Fukushima was the nightmare scenario that CNIC had long predicted. In a 1995 interview, Takagi spoke about the risks of a meltdown in the event of multiple failures. He raised the possibility of large radioactive releases from a meltdown resulting from a breakdown in the emergency core cooling system and the failure of backup diesel generators, exactly what happened at Fukushima sixteen years later (Takagi 1995).

Warnings by the CNIC and other anti-nuclear activists and experts were not taken seriously by the nuclear village since it would have required abandoning their quest for nuclear power under Japan's seismically perilous conditions. As Perrow argues, 'There is the problem that warnings are often seen as mere obstructionism. This was the view of a representative for a Japanese utility who brushed away the possibility that two backup electrical generators would fail simultaneously' (Perrow 2011, p. 48). This expert witness testified at the Shizuoka District Court in February 2007 on behalf of Chubu Electric Power Co., the utility that owns and operates the Hamaoka nuclear power plant. Exasperated by questioning from the plaintiff's lawyers concerning what would happen in the event of a station blackout and loss of all backup electricity (as happened at Fukushima four years later), this irritated witness blurted out, 'If we took all these possibilities into account, we could never build anything' (Repeta 2011). This witness was Haruki Madarame, who was subsequently named chairman of the government's five-member Nuclear Safety Commission in April 2010.

In Japan, cozy and collusive ties between regulators and industry embodied in the *amakudari* system and the nuclear village have compromised nuclear safety. *Amakudari* literally means descent from heaven, and in practice means officials securing post-retirement sinecures in the industry they previously supervised in their official capacity, creating a conflict of interest; officials are loath to alienate potential future employers by zealous enforcement of regulations and standards (*New York Times*, 26 April 2011). This situation has led to widespread regulatory capture, explaining the lack of a culture of safety at TEPCO and the averted eyes approach to monitoring the nuclear industry evident at NISA. Workers at Fukushima report being routinely warned in advance of inspections and inspectors did not seem eager to uncover violations.

Former PM Noda once stated that he did not support building any new reactors, did not favor extending the operating licenses of aging plants beyond their original life spans and supported a gradual phasing out of nuclear energy. He later backtracked from these positions. In particular, he was much more favorably inclined towards nuclear energy than his predecessor Naoto Kan. In July 2011 Kan called for the gradual phasing out of nuclear energy, stating that he believed it was not possible to operate nuclear reactors safely in Japan. In contrast, Noda restarted reactors and stressed the importance of nuclear energy to Japan's economy. His Cabinet also approved the completion of three reactors under construction and introduced legislation which allows the extension of operating licenses for aging reactors (> 40 years) (Kingston 2012c).

This new Japanese law requires the decommissioning of aging plants, but features a critical loophole designed to permit their continued operation at the discretion of regulators. Given the track record of regulators in Japan (and the US), what is supposed to happen only in exceptional cases (continued operation of old reactors) may become the norm. In fact, in June 2012 NISA gave provisional approval to the extension of the Mihama plant's operating license even though two of the reactors are over 40 years old and the other is over 35 years old. Given that so many of Japan's reactors are aging (three are over 40 years old and another 16 are over 30 years old), with the attendant risks of metal fatigue and dated technology, safety issues are becoming ever more urgent; the three meltdowns at Fukushima occurred in reactors commissioned in 1971, 1974 and 1976. Policymakers, under the pretext of mandating decommissioning such aging reactors, have actually ensured that the government retains discretionary powers to extend operating licenses and have also lengthened the timeframes in between inspections, in an effort to improve the lifetime profitability of all reactors (Sawa 2012). These initiatives are increasing the risk of operating nuclear reactors.

Whistleblower revelations of systematic falsification of repair and maintenance records in 2000 at all of TEPCO's nuclear plants indicate that more robust inspections, transparency and accountability are crucial to nurture a culture of safety (Kingston 2011a, pp. 151–2). It is indicative that in February 2011, shortly before the meltdowns, NISA extended the operating license of Fukushima Dai-ichi despite expressing reservations about a dubious maintenance record and eerily prescient concerns about stress cracks in the backup diesel generators that left them vulnerable to inundation. Former PM Noda also shortchanged safety by hastily pushing through the decision to restart reactors even though the Oi reactors only met 20 out of the 30 safety guidelines drawn up by his cabinet in May 2012. The lessons of Fukushima and the dangers of cutting costs on safety measures are not being heeded, raising the risks to human security.

Shifting blame

So who is to blame for the three meltdowns at Fukushima? The nuclear village tried to shift blame onto PM Naoto Kan, spreading erroneous information about his visit to Fukushima Dai-ichi forcing TEPCO to stop venting and subsequently alleging that he ordered the halt of pumping of seawater to cool the reactors and spent fuel rods stored in adjacent pools (Kingston 2012a). The failure to vent did lead to hydrogen explosions in three secondary containment buildings, but this was TEPCO's responsibility and had nothing to do with Kan's visit on 12 March (Kingston 2011c). Similarly, PM Kan never ordered the cessation of seawater pumping and the plant manager actually ignored instructions from the TEPCO president to do so, because under international protocols it is his call. TEPCO retracted its allegations, but not before damaging Kan's reputation and sowing suspicions about his responsibility for the nuclear crisis. Scapegoating Kan served many purposes, especially diverting attention away from the responsibility of TEPCO, NISA and the Ministry of Economy, Trade and Industry (METI) for the accident and woeful crisis response. The LDP also needed political cover since it was the party in power that had promoted nuclear energy and was complicit in the lax oversight that undermined plant safety. Personalizing the problem was an effort to downplay the fundamental institutional flaws that lay at the heart of the crisis. Discrediting Kan also served to discredit his anti-nuclear, pro-renewable energy initiatives.

To his credit, PM Kan in dealing with the disaster did not trust the responsible bureaucrats advising him or TEPCO. But Kan's justified suspicions also left him isolated and unable to call on people and institutions with relevant expertise. According to the *New York Times*, 'At the drama's heart was an outsider prime minister who saw the need for quick action but whose well-founded mistrust of a system of alliances between powerful plant operators, compliant bureaucrats and sympathetic politicians deprived Prime Minister Kan of resources he could have used to make better-informed decisions' (12 June 2011). As a result, those without expertise were making crucial decisions while experts such as NSC Chairman Madarame were giving misleading advice, inevitably leading to mistakes and zigzagging.

NISA was responsible for instituting government crisis procedures, and TEPCO was responsible for safe operations of its plants, but both were unprepared when it counted most. On 17 February 2012 former PM Kan asserted that Fukushima was a man-made disaster and that the authorities were woefully unprepared to deal with it (*Japan Times*, 18 February 2012). All three investigations into the nuclear accident cited above also highlight the role of human error in the Fukushima accident and reject TEPCO's assertion that it could not have been anticipated. The National Diet report finds:

> The earthquake and tsunami of March 11, 2011 were natural disasters of a magnitude that shocked the entire world. Although triggered by these cataclysmic events, the subsequent accident at the Fukushima Daiichi Nuclear Power Plant cannot be regarded as a natural disaster. It was a profoundly manmade disaster – that could and should have been foreseen and prevented. And its effects could have been mitigated by a more effective human response.
> (National Diet 2012, p. 9)

Kan further explained that there were no systems or procedures in place to respond effectively to Fukushima and officials had to make it up as they went along. 'Before 3/11, we were totally unprepared', he said. 'Not only in terms of the hardware, but our system and the organization were not prepared. That was the biggest problem.' Kan also acknowledged that information dissemination was slow and sometimes inaccurate, asserting that the disaster exposed a wide range of vulnerabilities and risks and the need to overhaul safety guidelines and improve crisis management. Three investigations into the Fukushima disaster, conducted by a government panel, a non-government panel and the Diet, criticized Kan for micromanaging and meddling in the crisis response at the nuclear plant and for closeting himself with a small coterie of trusted advisors. Only the non-government panel praised him for refusing TEPCO's requests on 15 March to abandon Fukushima Dai-ichi and ordering the utility not to withdraw its staff from the stricken plant (Funabashi 2012). However, all three investigations conclude that the institutions that should have been prepared to manage the crisis – TEPCO, METI, NISA and the NSC – were totally unprepared and thus not responding effectively.[1] Kan very quickly sensed this vacuum in the crisis response and was trying to compensate for the shortcomings of the responsible institutions. Thus to blame him for meddling seems to overlook the context of dysfunctional, paralyzed institutions and what was at stake if he shied away from intervening.

The Rebuild Japan Initiative Foundation (RJIF), the non-government investigation, interviewed all the people in the room with the premier, including those who were critical of his crisis management, when TEPCO made its request to evacuate personnel from Fukushima Dai-ichi and they all corroborated Kan's charge that TEPCO had proposed a total evacuation and repudiated TEPCO's subsequent assertions that it was not proposing to totally abandon the nuclear plant and planned to leave behind a skeleton crew (Funabashi Report 2012). TEPCO and its defenders blamed General Electric for the accident because it supplied the plant design right down to the placement of the backup generators and refused to modify it despite concerns expressed by local contractors at the time about the need to protect against tsunamis. TEPCO also conducted an in-house investigation that blamed PM Kan and asserted that the massive tsunami was an unforeseeable and unimaginable (*sotegai*) natural event, a claim that has been refuted (Kingston 2012b).

Crisis assessment

An investigation led by disaster expert Yotaro Hatamura about the nuclear crisis at the behest of the government issued an interim report at the end of 2011 that was harshly critical of TEPCO and the government (Hatamura 2012). This panel of experts pointed out that the utility was ill-prepared for a crisis and that its workers made critical errors in shutting off automated emergency cooling systems and wrongly assumed part of the cooling system was working when it was not (*Asahi Shimbun*, 27 Dec. 2011; *New York Times*, 26 Dec. 2011). These workers and their managers were inadequately trained to cope with an emergency situation and according to the panel lacked basic knowledge concerning the emergency reactor cooling system. Their mishandling of emergency procedures contributed to the crisis. Moreover, TEPCO and its regulators failed to act on fresh and compelling evidence about tsunami risk, a blind spot that left the plant needlessly vulnerable. Because the possibility of a tsunami inundating the plant was ignored, TEPCO made no preparations for simultaneous and multiple losses of power. In 1993 NISA had called on utilities to develop backup power systems in order to prevent a station blackout in the event of a tsunami, but the utilities lobbied regulators and did not institute such measures. On 3/11, the Fukushima station blackout that utilities claimed would never happen halted cooling systems, caused the meltdowns and disrupted communications among emergency workers and between the plant and the government. Meanwhile, the government was kept in the dark about critical developments and officials delayed giving advice to the Prime Minister and his advisors on how to respond to the nuclear crisis.

Investigators concluded that TEPCO failed to provide information to the government in a timely manner because it was inadequately prepared for an emergency. The crisis management center for Fukushima Dai-ichi was only 5 km from the plant, and when plant workers arrived they found it wrecked, with no power or functioning communications, and unusable because there was no system to filter out radiation. This poor emergency preparedness delayed the flow of information to the Prime Minister's office, slowing the government response. NISA was widely criticized for not having done more over the years to force TEPCO to improve its preventive and emergency measures. It was also revealed that NISA staff abandoned the Fukushima plant after the earthquake on 11 March and thus could not collect and disseminate live information as the crisis worsened; after being ordered to return, they did little to help manage the crisis.

All three investigations also pilloried TEPCO and the government's mishandling of the evacuation of residents living near the plant, in many instances evacuating people to places where levels of radiation were higher than those they had left. This reflected the general problem of information bottlenecks; PM Kan and his Cabinet were not given data on radiation contamination that could have led to a more sensible evacuation order; some designated evacuation sites were actually hot zones. The three investigations confirm that data generated by the System for Prediction of Environmental Emergency Dose Information (SPEEDI) on radiation dispersal was available and could have been used to evacuate residents at greatest risk to safer areas, but this information was not provided to the Prime Minister's crisis management center until 23 March, eleven days after the first hydrogen explosion released plumes of radioactive substances into the air. The US also provided radiation dispersal data to the Ministry of Foreign Affairs soon after the March meltdowns, but again this information remained in bureaucratic hands and was not shared with the premier. Finally, one month after the original evacuation, the government used this SPEEDI data to move evacuees out of harm's way, meaning that many had been subjected to substantial doses of avoidable radiation exposure. When officials responsible for SPEEDI were asked why they did not make this crucial data available to crisis managers sooner, they lamely replied that nobody asked them for it. Banri Kaieda, METI Minister during the crisis, and the top official responsible for the nuclear energy industry, admitted he had never even heard of the SPEEDI system before the accident. He also admitted that his office bore some of the responsibility for the two and a half hour delay in declaring a nuclear emergency on 3/11 after TEPCO informed METI about the station blackout and probability of meltdowns.

Tsunami risk

> It's inexcusable that a nuclear accident couldn't be managed because a major event such as the tsunami exceeded expectations.
>
> (Hatamura, NHK News, 26 Dec. 2011)

Yotaro Hatamura chaired an investigation into the Fukushima accident and is a well-known authority on accidents (Hatamura 2008). He has analyzed data on over 1,100 industrial accidents focusing on design flaws, system failures and human error. For Hatamura, managing risk at a nuclear power plant is about foreseeing the unforeseen and preparing accordingly. His committee refuted TEPCO's in-house, self-exonerating report that blamed the accident on an inconceivable, rare natural disaster. In fact, TEPCO ignored several warnings, including internal research, about the possibility of a monster tsunami. It looked into building a larger tsunami seawall, but decided the cost was prohibitive and took no additional preventive measures. On 7 March 2011, only four days before the tsunami, TEPCO actually presented NISA, the government nuclear watchdog authority, with results from simulations conducted in 2008 by its own researchers showing that a tsunami as high as 15.7 meters could hit the area, a finding it ignored.

Tell-tale warnings began accumulating over the decade prior to 3/11. In 2001, researchers cited geological evidence that the Jogan tsunami of 869 slammed the Fukushima coastline and the wave height was strikingly similar to the 3/11 event. Their research on ancient gigantic tsunamis noted that such uncommon disasters occur every 800–1,100 years and specifically warned that the region was overdue for another. In February 2002 the Japan Society of Civil Engineers using new simulation techniques determined that there was a risk of a 5.7 meter tsunami and a month later TEPCO increased its estimates accordingly from the original assumption of a 3.1 meter tsunami when the reactor was being built in the early 1970s. In July 2002 the government's Headquarters for Earthquake Research Promotion warned that an even larger tsunami was possible based on historical evidence. In 2006 the government revised its anti-seismic guidelines, specifically calling on utilities to prepare for rare events. In 2009 NISA and TEPCO discussed the possibility of a 9.2 meter tsunami, based on new simulations and archaeological evidence, but NISA did not press TEPCO to take countermeasures and the utility judged that a higher seawall was prohibitively expensive.

Clearly, there is no basis to TEPCO's claim that the scale of the 3/11 tsunami was inconceivable; the utility chose to ignore centuries of geological evidence and repeated twenty-first-century warnings from modern scientists, including in-house researchers. In terms of tsunami-related risk management, it turns out that

TEPCO and two other utilities actually lobbied the government's Earthquake Research Committee on 3 March 2011 to tone down wording in a report warning that a massive tsunami could hit the Tōhoku coast (*Japan Times*, 27 Feb. 2012). The Fukushima nuclear reactors were vulnerable because TEPCO did not approach risk assessment from the basis of a worst-case scenario, making unduly optimistic assumptions that wished away a cataclysm in a region with a history of killer waves. As the UNDP noted in 1994, prevention is the most cost-effective approach to disaster management, but in Fukushima TEPCO did not want to spend the money necessary to protect against a large tsunami. This unjustified insouciance and penurious approach cost Japan dearly; TEPCO's bottom line mattered more than public safety. Prioritizing private assets over the public interest is yet another example of how human security was shortchanged in Fukushima due to regulatory capture, where regulators regulated in favor of the regulated.

Culture of safety?

Inexcusably, TEPCO did not make safety its ethos, while lax oversight by the government allowed this culture of complacency to persist long after it was obvious that TEPCO was cutting corners to cut costs. METI did shut down all 17 of TEPCO's reactors in 2002, but only because the media reported a whistleblower's revelations about systematic falsification of repair and maintenance records, and exposed the government's initial failure to act on this information. Investigations reveal that TEPCO's safety precautions were based on unrealistic assumptions that left the utility poorly prepared to deal with a crisis, a finding that came too late for the people evacuated from their homes in Fukushima and thousands of farmers and fishermen who lost their livelihoods.

Given the risks associated with operating nuclear power plants in a seismically active, densely populated country it is extraordinary that Japan's utilities did not practice evacuation procedures in reactor-hosting communities. The utilities justify this oversight by arguing that they did not want to alarm local residents by practicing for an unlikely event and thereby undermine repeated assurances that nuclear energy is completely safe. The lack of procedures and guidelines proved a major hole in disaster preparedness. In retrospect, this policy of preserving the myth of 100 per cent safety at the expense of actually safeguarding residents represents an institutionalized inclination to collectively 'bury heads in the sand' and irresponsibly minimize risk in ways that endangered local residents.

Transparency

In August 2011 a Diet committee investigating the nuclear disaster requested that TEPCO provide it with an operations manual for the Fukushima Dai-ichi plant. TEPCO initially refused the request, prompting a public uproar. One month later, TEPCO provided a heavily redacted version of the manual and justified blacking out key passages related to emergency procedures, arguing that this information constituted intellectual property it wished to protect and also raised security concerns. These spurious grounds highlighted TEPCO efforts to prevent the Diet from exercising oversight and to cover up shortcomings in its crisis response. It took six months for TEPCO to release the entire manual. Committee members complained about this stonewalling and stated that, 'It was important that we saw the manual to learn why the company had switched part of the emergency core-cooling system off and on again after the earthquake (and before the tsunami) – to find out when the emergency systems were destroyed' (Taira and Hatoyama 2011). But it was not only TEPCO that was attempting to cover its tracks.

In January 2012 the media reported that various government panels dealing with the Fukushima crisis failed to keep minutes of the proceedings, including the task force set up by the Prime Minister's office. Keeping minutes is standard procedure for government panels, one usually carried out by bureaucratic officials. The failure to keep minutes is a critical oversight because it prevents learning more lessons about the crisis response to avoid repeating the same mistakes in the future. *Asahi Shimbun* termed this absence of minutes a 'monumental level of government ineptitude', fuming that,

> It would be hard for the officials involved to disprove the charge that they deliberately neglected to keep a record of the meetings so that their blunders and missteps would not come to light later. Now, the oft-repeated pledge by top government officials to share lessons learned by the accident with the international community sounds hollow. Technically, the responsibility for this fiasco lies with the Nuclear and Industrial Safety Agency of the Ministry of Economy, Trade and Industry, which served as the secretariat for the headquarters. But even more to blame are the politicians who failed to ensure that a record of the meetings would be kept.
>
> (26 Jan. 2012)

Apparently, managing risk was more a matter of concealing chaotic and inconsistent decision-making and inadequate crisis response procedures than gleaning useful lessons about how to improve crisis response mechanisms.

Finally on 9 March 2012 the government released a 76-page summary of the 23 meetings of the Prime Minister's crisis management team that was reconstructed from interviews conducted in early 2012 with officials attending the 2011 meetings and unofficial notes kept by NISA officials. NHK contrasted this post-facto summary, one short on details, with the release of the 3,200-page transcript of the US Nuclear Regulatory Commission crisis deliberations about Fukushima. The summary is vague on important issues such as the decision to declare a nuclear emergency, discussions about a meltdown within hours of the earthquake, the decisions to expand the evacuation zone and criticism about the lack of a chain of command in managing the crisis. This belated attempt to quell public concerns about the lack of transparency actually amplified them. As a result, in June 2012 the government formally required all panels established in connection with an emergency to record minutes and produce summaries.

Whistleblower

In Diet testimony on 15 February 2012, Haruki Madarame, chairman of the Nuclear Safety Commission, pulled back the curtain on the nuclear village, drawing attention to cozy and collusive relations between regulators and the utilities, and lax safety standards. He spoke of officials ignoring nuclear risks and admitted, 'We ended up wasting our time looking for excuses that these measures are not needed in Japan' (Associated Press, 15 Feb. 2012). He asserted that Japan's safety monitoring technology was three decades out of date, while acknowledging that he and his colleagues had 'succumbed to a blind belief in the country's technical prowess and failed to thoroughly assess the risks of building nuclear reactors in an earthquake-prone country' (*New York Times*, 15 Feb. 2012; *Bloomberg*, 16 Feb. 2012). He said that regulators and the utilities missed many opportunities to improve operating safety and warned that safety regulations were minimally enforced and fundamentally flawed. Furthermore, he asserted, regulators were toothless and overly solicitous of utility interests. He acknowledged that officials did not prepare for a simultaneous station blackout and failure of backup generators and ignored a series of warnings about the dangers of a large tsunami affecting the Fukushima plant.

In Madarame's view, nuclear reactor safety had been compromised because of institutional complacency and perfunctory enforcement of safety regulations and guidelines. It was unnerving to have one of the nation's leading nuclear energy experts, the man in charge of the NSC, one who had long been a stalwart advocate of nuclear energy and who defended the nuclear crisis response in the months following 3/11, suddenly voice many of the same objections that anti-nuclear activists had expressed over the years. Madarame as apostate may not be convincing, but his withering indictment of the nuclear power industry and government regulators is an astonishing development in the post-Fukushima discourse. Of course some of it can be attributed to his desire to restore a battered reputation and to shift responsibility, as a result of which he has exposed the shadowy practices of the nuclear village, including collective heedlessness about safety and poor risk management. In the one sector where a culture of safety should have been foremost, the nuclear safety czar revealed a culture of deceit.

Shortly after his Diet testimony, Madarame dropped another bombshell when he announced that he did not think that the first round of stress tests conducted on Japan's nuclear reactors were sufficient to ensure safe operation. Speaking on behalf of the NSC, Madarame said, 'With only the first round (of stress tests), the level of safety confirmation that the commission seeks would not be met' (*Mainichi*, 21 Feb. 2012). This high-profile indictment of the stress tests came at an inconvenient time for the government because NISA had already endorsed first-stage stress tests conducted on Kansai Electric's Oi power plant. Nothing could be quite as damning as the NSC chairman, one of the nuclear village's headmen, pointedly refusing to endorse the stress tests.

The stress tests were first announced by PM Kan in July 2011, stirring considerable controversy because he did not consult with his Cabinet beforehand (Kingston 2012a, pp. 194–6). Kan's insistence on EU-style two-stage stress tests derailed METI's plans to quickly restart idle reactors and infuriated METI Minister Kaieda. METI had engaged in a PR campaign to reassure hosting communities that reactors were safe, and on 18 June 2011 Kaieda announced that METI had confirmed it was safe to resume operation of the nation's reactors. But this haste to resume business as usual in the nuclear industry only three months after the three meltdowns, and not quite one month after TEPCO finally admitted to the meltdowns, backfired. The media exposed how NISA and Kyushu Electric, at the suggestion of the governor of Saga, had orchestrated an internet 'town hall' meeting on 26 June, planting questions and opinions among participating 'netizens' in favor of nuclear energy and restarting the Genkai reactors in Saga Prefecture (*Yomiuri,* 14 July 2011; NHK TV News, 14 July 2011; *Japan Times,* 4 Aug. 2011). METI's plans suddenly came under fire and Kan seized the opportunity to introduce stress tests and handed responsibility for overseeing the process to NISA and the NSC.

At the time it looked like little more than a delaying measure because the utilities would conduct the computer simulations about the safety of restarting their own idled reactors. Adding to the conflict of interest, key institutions in the nuclear village, NISA and the NSC, were given responsibility for assessing the results. But the NSC upset these plans and in doing so stoked public skepticism about the effectiveness of stress tests based solely on computer simulations. Experts have pointed out numerous flaws in the stress tests and note that they do not measure metal fatigue, an important issue for aging plants, don't examine multiple failures as occurred at Fukushima and lack hands-on testing of components. The public also have good reason not to trust the utilities to report inconvenient findings from the stress tests since they have falsified repair and maintenance records. Moreover, TEPCO conducted computer simulations in 2008 on massive tsunamis that it failed to share with NISA until four days before 3/11. Critics also point out that much depends on what assumptions are used in the simulations and doubt that utilities, with so much investment in nuclear energy at stake, will uncover the need for expensive retrofitting or decommissioning.

Madarame undermined the credibility of the stress tests and indicated that more sweeping reforms are needed to upgrade safety and monitoring in the nuclear industry, but PM Noda was undeterred. The government vaguely promised to take into account the feelings of local residents while rejecting calls for a national referendum. Instead, the government only consulted residents of Oi, a town that depends on the economic benefits of hosting a nuclear plant, who are paid only when it is running. This model of consultation suggests that the government plans to seek approval for reactor restarts only from those who have the most to gain from resuming operations, based on tests conducted by those with the most to gain from going back online. This strategy limits the risk of confronting anti-nuclear public opinion, but Fukushima has changed perceptions about nuclear power safety.

For example, an NHK poll in October 2011 indicated that 80 per cent of the mayors of hosting communities oppose restarting reactors until safety can be verified. The governor of Niigata, which hosts the massive Kashiwazaki nuclear plant, has repeatedly stated that he would oppose resumption of operations until the Fukushima crisis is resolved and dismisses the value of the stress tests. So it may well be true that local people can be convinced or bribed into restarting idled reactors, but they are also keenly aware of how little has been done for the residents of Fukushima and how much they lost.

In Fukushima evacuation centers there is a degree of tension between evacuees from hosting villages and those from neighboring villages that never received any subsidies or benefits, but have experienced the same level of personal loss and dislocation (Horner 2012). The politics of cherry-picking public opinion are uncertain, but the government does seem to be courting risk by ignoring the voices of many other local residents who have just as much at risk as hosting community residents and nothing to gain.

More worrisome for the nuclear village is the March 2012 NHK poll conducted in 142 communities in the vicinity of Japan's nuclear power plants. NHK found that only 14 per cent of respondents favored restarting idled reactors now or in the near future while 79 per cent opposed or had strong reservations about doing so (NHK News, 8 March 2012). Clearly the government faces a steep uphill battle in gaining the understanding of Japanese living near nuclear reactors about plans for restarting reactors.

Backlash

Over the past five decades the government and utilities have educated Japanese citizens to believe in the safety, reliability and necessity of nuclear energy. Indeed this myth became self-deluding, blinding regulators and operators to the risks and rendering adequate crisis management procedures taboo. Thus the Fukushima debacle came as a shock to most Japanese, one that thoroughly undermined the assiduously propagated myths of nuclear energy safety. As the crisis lingered and it became clear that the nation would be dealing with the consequences for decades rather than months, shock gave way to a backlash, jolting many citizens out of resignation to varying degrees of anti-nuclear activism (Liscutin 2011; Oka Norimatsu 2011).

The top-down consensus promoting nuclear energy is now being challenged by a growing bottom-up backlash, fueled by public concerns that the government is ignoring the lessons of Fukushima in ways that continue to undermine human security. From the end of March 2012, on every Friday anti-nuclear demonstrators gathered in front of the Prime Minister's office. Initially these gatherings were sparsely attended, but following the decision to restart the Oi reactors the crowds grew from hundreds to thousands into the tens of thousands. Ordinary citizens have been provoked by the government's hasty decision, one that was based on ad hoc criteria and sidelined public opinion. The summer of 2012 witnessed the largest demonstrations in Tokyo since the 1960s. Significantly, widespread anxieties and mistrust are sabotaging government efforts to resume business as usual in the nuclear industry. This activism is also evident in social media where websites post radiation readings taken by 'citizen scientists' armed with their own Geiger counters that map the spread and extent of contamination, painting a far more grisly situation than official assessments. In addition, there is also a citizen's campaign led by Nobel Literature laureate Oe Kenzaburo, among others, to collect 10 million signatures for an anti-nuclear energy petition; as of October 2013 they had over 8 million.

In this siege, the moat surrounding the nuclear village has been breached, but the ramparts remain well defended. Activists have sought a national referendum on nuclear energy and various local referenda are being mooted, but the government has a track record of ignoring such plebiscites (Numata 2006). The popular mayor of Osaka, Hashimoto Toru, has been very critical of the utilities and nuclear energy, despite the fact that it previously provided 50 per cent of electricity in Kansai. He too, however, has derailed a referendum initiative.

Citizens are also responding to the nuclear crisis through voluntary conservation efforts. In the summer of 2011 there were expectations of rolling blackouts as reactors went offline for regular inspections, so the government mandated conservation for large commercial users and urged the public to reduce electricity consumption by 15 per cent. Through lifestyle changes and innovative measures, the public exceeded this target and registered an overall 20 per cent decrease in electricity consumption. Surveys indicate that some 60 per cent of the public have practiced conservation since 3/11 and it seems to be a new commonsense norm, one that will be reinforced by higher electricity prices. Polls also indicate strong public support for renewable energy (DeWit 2011). In the record-breaking heat-wave in the summer of 2012, there were no blackouts and there was no need for the electricity generated by the Oi nuclear reactors. Even without mandatory cuts in power usage and not much in the way of a public campaign encouraging conservation (in contrast to 2011), overall electricity consumption declined; in metropolitan Tokyo demand dropped 16 per cent in July 2012 compared to July 2010.

Conclusion

The nuclear crisis at Fukushima was triggered by natural disaster, but human error played a critical role. A systemic failure in risk management, institutionalized complacency about tsunami risk and incompetence in operating emergency cooling systems were crucial factors in this catastrophe. TEPCO lacked a culture of safety that explains its lapses before, during and after 3/11. Fukushima was an accident waiting to happen and nuclear industry regulatory authorities are complicit because they failed to pressure TEPCO to heed numerous warning signs. Because risks were downplayed, TEPCO and the government were poorly prepared to deal with the meltdowns and respond effectively to the consequences of the accident. It is equally alarming to know that the scientific community did little to challenge, and in the end perpetuated, the absolute safety myth that enshrouded nuclear energy.

The mishandling of the evacuation subjected many Fukushima residents needlessly to radioactive contamination, highlighting authorities' inadequate planning for a nuclear crisis. Other communities hosting nuclear plants have taken note of lax disaster preparedness and how little has been done for the Fukushima evacuees. As a result, restarting reactors shut down for inspections and stress tests is proving politically divisive. Decontamination, decommissioning and disposing of contaminated waste over the coming decades will keep nuclear energy under sustained, critical scrutiny. For many Japanese, Fukushima demonstrated just how vulnerable human security is in Japan precisely because the government has promoted nuclear energy without enacting robust safety standards and even made excuses why it need not adapt international standards. Fukushima is an example of how the three phases of disaster risk management – prevention, response and rebuilding – are interrelated. 'Good' prevention is a solution multiplier in the response phase, and 'bad' prevention is a threat multiplier in the response phase. In Fukushima, prevention was constrained by the taboo of suggesting that there might be a risk associated with nuclear energy and the costs of mitigating such risks. In the nuclear industry, regulators and operators were blinded to the risks because they perpetuated the self-deluding myth of 100 per cent safety and as a result did not prepare for worst-case scenarios. This self-delusion among vested interests highlights a significant vulnerability in emergency preparedness and disaster risk management where people who should be asking tough questions and imagining the inconceivable become prisoners of their own propaganda. Institutionalized complacency was and continues to be a major challenge.

TEPCO's risk management prior to and during the crisis may have been woeful, but in the aftermath it has been relatively successful in managing risks to its institutional interests and avoiding accountability. While its reputation may be in tatters, TEPCO has stonewalled ceding management power to the government while obtaining vast sums of public money to cover the utility's enormous costs for clean-up, disposal, decontamination, decommissioning and compensation. It has also attempted to block attempts to separate power generation from distribution, in a bid to maintain advantages that would impede the expansion of renewable energy capacity. In March 2012, a year after the Japanese triple disaster, *The Economist* issued a Special Report on nuclear energy titled 'The dream that failed', opining that 'they allowed their enthusiasm for nuclear power to shelter weak regulation, safety systems that failed to work and a culpable ignorance of the tectonic risks the reactors faced, all the while blithely promulgating a myth of nuclear safety'. The great risk in Japan today and well into the future is that the lessons of Fukushima are being skewed, ignored or marginalized in a nation where nuclear energy represents a significant and abiding risk to human security.

Note

1 Tokyo Electric Power Company (TEPCO), operated the Fukushima nuclear plant. The Ministry of Economy, Trade and Industry (METI) is responsible for overseeing the nuclear industry and has long been an advocate of nuclear energy. The Nuclear Industrial Safety Agency (NISA) operated from within METI and was directly responsible for regulating the nuclear power industry. As of September 2012 it was disbanded and replaced by the Nuclear Regulatory Authority (NRA). The Nuclear Safety Commission (NSC) was the Cabinet nuclear safety advisory board and reviewed actions by nuclear regulatory agencies. It has now been merged into the NRA.

References

Aldrich, D. 2008, *Site Fights: Divisive Facilities and Civil Society in Japan and the West*, Ithaca, NY: Cornell University Press.

Avenell, S. 2012, 'From fearsome pollution to Fukushima: Environmental activism and the nuclear blind spot in contemporary Japan', *Environmental History*, 17, April, pp. 244–276.

Cabinet Office. 2011, *Japan: National Progress Report on Implementation of the Hyogo Framework for Action (2009–2011)*, Tokyo: Government of Japan – International Office of Disaster Management, 25 April.

DeWit, A. 2011, 'Fallout from the Fukushima shock: Japan's emerging energy policy', *Asia-Pacific Journal*, 9/45(5), 7 Nov.

Funabashi, Y. 2012, Press conference, Foreign Correspondents Club, Japan, 1 March.

Funabashi Report. 2012, *The Independent Investigation Commission on the Fukushima Nuclear Accident*, Tokyo: Rebuild Japan Initiative Foundation, http://rebuildjpn.org/en/fukushima/report, viewed Dec. 2013.

Hatamura, Y. 2008, *Learning from Design Failures*, New York: Springer.

Hatamura Report. 2012, *Investigation Committee on the Accident at the Fukushima Nuclear Power Stations of TEPCO*, http://icanps.go.jp/eng, viewed Dec. 2013.

Horner, N. 2012, Correspondent for Swedish Broadcasting Corporation, personal communication, 2 Feb.

Kingston, J. 2011a, *Contemporary Japan: History, Politics and Social Change since the 1980s*, Chichester: Wiley.

Kingston, J. 2011b, *Tsunami: Japan's Post-Fukushima Future*, Washington, DC: Foreign Policy.

Kingston, J. 2011c, 'Ousting Kan: The politics of nuclear crisis and renewable energy in Japan', *Asia-Pacific Journal*, 9/39(5), 26 Sept.

Kingston, J. 2012a, 'The politics of disaster, nuclear crisis and recovery', in J. Kingston (ed.), *Natural Disaster and Nuclear Crisis in Japan: Response and Recovery After Japan's 3/11*, New York: Routledge, pp. 188–206.

Kingston, J. 2012b, 'Mismanaging risk and the Fukushima nuclear crisis', *Asia-Pacific Journal*, 10/12(4), 19 March.

Kingston, J. 2012c, 'Japan's Nuclear Village', *Asia-Pacific Journal*, 10/37(1), 10 Sept.

Liscutin, N. 2011, 'Indignez-vous! "Fukushima", new media and anti-nuclear activism in Japan', *Asia-Pacific Journal*, 9/47(1), 21 Nov.

National Diet of Japan. 2012, *Official Report*, Tokyo: Fukushima Nuclear Accident Independent Investigation Commission, 5 July, http://naiic.go.jp/en, viewed Dec. 2013.

Numata, C. 2006, 'Checking the center: Popular Referenda in Japan', *Social Science Japan Journal*, 9(1), pp. 19–31.

Oka Norimatsu, S. 2011, 'Fukushima and Okinawa – the "abandoned people", and civic empowerment', *Asia-Pacific Journal*, 9/47(3), 21 Nov.

Onitsuka, H. 2012, 'Hooked on nuclear power: Japanese state–local relations and the vicious cycle of nuclear dependence', *Asia-Pacific Journal*, 10/3(1), 16 Jan.

Perrow, C. 2011, 'Fukushima and the inevitability of accidents', *Bulletin of the Atomic Scientists*, 67(6), pp. 44–52.

Repeta, L. 2011, 'Could the meltdown have been avoided?', in J. Kingston (ed.), *Tsunami: Japan's Post-Fukushima Future*, Washington, DC: Foreign Policy, pp. 183–94.

Sawa, T. 2012, 'Tradeoff in nuclear power', *Japan Times*, 27 Feb.

Taira, T. and Hatoyama, Y. 2011, 'Nuclear energy: Nationalize the Fukushima Daiichi atomic plant', *Nature*, 480, 14 Dec., pp. 313–14.

Takagi, J. 1995, 'Kakushisetsu to Hijojitai: Jishin Taisaku no Kensho o chushin ni', *Nihonbutsuri Gakkaishi*, 50: 821.

The Economist 2012, 'Nuclear power: The dream that failed', 8 March.

4 Hidden insecurities

The workers of Fukushima Dai-ichi

Christopher Hobson

Introduction

When surveying the wreckage left behind by the massive tsunami that struck on 11 March 2011, some of the greatest damage could not be seen: it was the radiation escaping into the atmosphere from the failing Fukushima Dai-ichi power plant. Whereas the tsunami left more than 15,000 dead and 2,000 missing, the Fukushima meltdowns have yet to *directly* cause any deaths.[1] Nonetheless, the nuclear accident has proven to be just as traumatic, perhaps even more so due to its unfamiliar and ongoing nature. The horrible black waters that took so many lives on 3/11 quickly receded, but three years later the nuclear disaster continues. Many of the approximately 154,000 residents that remain displaced do not know when, if ever, they can return home. And even if they eventually can go back, it is unclear what kind of life will be waiting for them. If one recalls that human security is about creating and maintaining conditions in which people can have a certain level of regularity and normality in their lives, to live in a situation in which they are free from fear, then it is clear that the nuclear accident at Dai-ichi has generated considerable human insecurity.

When considering the human impact of the nuclear accident, understandably the focus has been on those who have lost their homes and livelihoods, and especially how it has affected mothers and their children. One group of people that has received less attention are the workers of Dai-ichi, who have toiled in incredibly difficult conditions. Considering that TEPCO has been unambiguously identified as the prime culprit for the accident, it is hardly surprising that there has been limited concern for the welfare of the people they employ. As Seiko Takahashi, a Fukushima mother who has become an anti-nuclear campaigner, explains, 'we see them as one block, they work for TEPCO, they earned high salaries' (Wingfield-Hayes 2013). This reflects a strong tendency in Japan to associate people with the company they work for. Those at Dai-ichi are identified as part of TEPCO, which means they belong to the company that is culpable for the second worst nuclear accident in human history. From this perspective, all the employees of TEPCO – from senior management down to the plant workers – are collectively perpetrators of this catastrophe. While such a perception is certainly understandable, it ignores the fact that a majority of those at Dai-ichi are contract workers who do not receive high salaries or generous benefits. Furthermore, many of the workers were also victims of the tsunami and nuclear accident, losing loved ones and possessions on 3/11. Indeed, they have much more in common with other victims of the nuclear accident than they do with TEPCO management. By disaggregating the TEPCO workforce, it can be further seen that those at the bottom – labourers employed at Dai-ichi through subcontractors – are acutely vulnerable. From the perspective of human security – which is concerned about the well-being of *all* people – it is vital to recognize and respond to the considerable hardships these workers face.

The purpose of this chapter is to examine the vulnerabilities that shape the lives of the workers at Fukushima Dai-ichi. To do so it is necessary to place the nuclear accident in a wider socio-economic context. Human security did not suddenly become a problem in Fukushima on 11 March 2011. Rather, existing insecurities were considerably exacerbated and new ones created. The majority of those employed at Dai-ichi were pushed into this line of work out of economic necessity. And when the tsunami struck, they were not immune from its wrath. Many lost family, friends and colleagues, had their homes destroyed and were forced to evacuate. While suffering from these traumatic experiences, they had to continue working at Dai-ichi in incredibly challenging circumstances as three of its reactors went into meltdown. The chapter considers the impact these difficult conditions have had on the workers, focusing on the areas of economic and health security. Working for limited wages and with few benefits, those employed to decommission Dai-ichi must manage in difficult conditions, with little sympathy or support because of their association with TEPCO. It is suggested that there is a need for their insecurities to be uncovered and better addressed. In concluding, the chapter will reflect on the long-term prospects for those affected by the Fukushima disaster.

Human insecurity before the triple disaster

Japan may have been one of the leading proponents of human security internationally, but prior to the triple disaster there was little realization that the approach was also applicable in a domestic context (see Chapter 2, by Bacon). Certainly what human insecurity looks like differs significantly from one part of the world to another, but one can still find serious instances of major insecurity existing even in wealthy, stable democracies. Looking at Japan, the way the nuclear energy industry developed was closely connected to significant socio-economic vulnerabilities in society. Indeed, it would not be a great overstatement to suggest that the Japanese nuclear industry has been partly built on exploiting vulnerable individuals and communities. Put in the language of human security, utilities like TEPCO preyed on 'economic' and 'community' insecurities, and this was taking place long before the events of 11 March 2011.

Fukushima's hosting of nuclear power plants reflected the relative weakness of the prefecture. It is worth recalling that all of the electricity generated by the Dai-ichi plant went to Tokyo, which outsourced the risk to the more isolated and poorer Fukushima region. This reflects the highly political nature of the siting of nuclear power plants, with most being built in rural, isolated areas suffering from shrinking populations, dwindling economies and fragmenting communal ties. Hosting nuclear plants offered much-needed sources of revenue and employment, leaving weak communities with few options but to accept. This 'free' choice was further limited in many cases by the hosting agreement being secured with little community input. The necessary land was bought and arrangements were made in advance, which meant few residents had a say in the decision-making process. In the case of Ōkuma, one of the closest towns to Fukushima Dai-ichi, a secret special committee was formed to prevent information relating to the possible plant becoming public (Onitsuka 2011). As a result, residents did not find out about the impending plant until two years *after* the town assembly had already approved it. This corresponds with Aldrich's (2008) argument that the quality and quantity of social capital have played an important role in determining which communities in Japan were more likely to become hosts of nuclear facilities. A combination of being socio-economically vulnerable, lacking information about possible risks, not being privy to secret discussions, and fragmented civil societies without the cohesiveness to organize opposition, meant that these communities only had the freedom to choose whether to accept nuclear facilities in a rather limited sense. If we recall that human security is about expanding 'the real freedoms that people enjoy' (CHS 2003, p. 10), this was certainly not how nuclear power developed in Japan.

It was not only the communities hosting the plants that were vulnerable. The nuclear industry has been built on a highly exploitative labour system that takes advantage of the socio-economically vulnerable position of most of its employees, whilst exposing them to considerable health and safety risks. In the language of human security, there have not only been problems of 'community security', but also 'economic security' and 'health security'. While most anti-nuclear protestors understandably associate anyone who works at Dai-ichi with TEPCO, in reality only a small proportion – the most elite and well educated – are directly employed by the power companies. In 2010, at Dai-ichi 89 per cent of the workers were contractors, almost exactly the same as the industry as a whole, in which 88 per cent of the approximately 83,000 nuclear workers in Japan were contractors (Tabuchi 2011a). On average a TEPCO employee would receive $94,000 per year, more than three times than most contractors, who were often paid under $30,000 per year (Tabuchi 2011b).

A defining characteristic of the Japanese nuclear industry is a complex system of subcontracting in which many workers are employed by organizations at five or six removes from the actual plant operator. At each level a cut of their pay is taken by the subcontractor, meaning that the actual wages received are significantly reduced. Through this process, workers are also deprived of important rights, often receiving insufficient health and safety training and poor-quality protective gear. There is a clear hierarchy, with employees of the energy companies undertaking the safest duties, while those at the bottom – the newest people subcontracted at the furthest remove – receiving the most dangerous tasks. This 'radiation fodder' is recruited from acutely vulnerable groups like the urban homeless (Jobin 2011, p. 37). This subcontracting system is stretched to its limits when reactors are taken offline for maintenance each year. At this time more workers are required and more dangerous tasks need to be undertaken. In some cases, the *yakuza* (Japanese mafia), who are often involved in the lower layers of subcontracting, have been known to use strong-arm tactics to force people to work (Adelstein 2012).

This subcontracting system undermines the rights and protections of workers and has resulted in greater exposure to health risks. For all nuclear workers there are limits on how much radiation they can be exposed to. Once the regulated amount has been exceeded, they are no longer supposed to continue working. Yet given that most have few other employment opportunities, there is a perverse incentive to not properly record their exposure levels so that they can keep on working. This has also resulted in a class of men known as 'nuclear gypsies', who travel from plant to plant to continue working, as there is no nation-wide database that compiles the total radiation exposure. While this allows them to continue being employed in the short to medium term, there are serious long-term ramifications: not only does it increase the chances of these workers getting cancer or another work-related illness; it also makes it more difficult for them to receive compensation if this happens, as they are lacking accurate documentation of their exposure levels. This is evidenced in the fact that in the long history of nuclear power in Japan there have been just 14 officially recognized cases of workplace disease (Jobin 2011, p. 38). The system of nuclear subcontracting illustrates the interconnected nature of human security threats. For the workers 'economic insecurity' in turn creates greater degrees of 'health insecurity'. France followed Japan in adopting this exploitative system of subcontracting, replicating a condition that Thébaud-Mony (2011) has provocatively termed 'nuclear servitude'.

At both an individual and societal level the nuclear industry in Japan has been predicated on, to a certain extent, the existence and exploitation of human insecurities. Weak and vulnerable communities are needed to host the plants, weak and vulnerable men are needed to run the plants. The interests of the so-called 'nuclear village' – the informal coalition of government, bureaucratic and industry forces aligned around nuclear power – have been constantly prioritized at the expense of the well-being of others. The key point here is that many of the issues that have been identified in the wake of the nuclear accident – weak, depopulating communities with few job opportunities; nuclear plants staffed by vulnerable men who must endure dangerous and exploitative conditions – existed in Fukushima, and other host communities, *before* the earthquake and tsunami struck. These problems were hidden in plain sight. As Hecht (2012) observes, 'systemic invisibility permeates the nuclear industry in both ordinary and extraordinary times'. What the nuclear accident did was expose and greatly exacerbate these pre-existing insecurities, while creating many new ones.

Human security in meltdown

When the earthquake and tsunami struck, the Fukushima Dai-ichi power plant was over 40 years old and one of the oldest operating in the country, which left it much more vulnerable to a major natural hazard. The Kurokawa Report observes that:

> The construction of the Fukushima Daiichi Plant that began in 1967 was based on the seismological knowledge at that time. As research continued over the years, researchers repeatedly pointed out the high possibility of tsunami levels reaching beyond the assumptions made at the time of construction, as well as the possibility of core damage in the case of such a tsunami. TEPCO overlooked these warnings, and the small margins of safety that existed were far from adequate for such an emergency situation.
>
> (NAIIC 2012, p. 27)

This left Dai-ichi ill-prepared when the massive earthquake and tsunami struck. The plant soon lost all power, including its backup generators, which meant there was no way to cool the reactors.[2] The government inspectors fled immediately after the tsunami, the TEPCO president was incommunicado, the plant had been seriously damaged and 'even worse, there was no plan for what to do next because nobody in TEPCO had ever predicted total loss of power at a nuclear plant' (Birmingham and McNeill 2012, p. 64). The Herculean task of stabilizing the failing plant and avoiding a far greater catastrophe was left to a skeleton crew of workers who came to be known as the 'Fukushima Fifty', even though the actual number was in the hundreds.

Conditions at Dai-ichi were very challenging for the workers, who had to do long shifts without proper light or ventilation. They slept on the floor and without blankets in a structure near the reactors with hourly radiation readings equivalent to that of a dental x-ray. Their meals were rationed, with biscuits and juice in the morning, and emergency rice rations and tinned meat at night (Penney 2011). Only seven workers perished, none because of radiation poisoning, but because they were forced to work without the proper safety equipment, as most of the radiation monitors were swept away or rendered unusable by the tsunami (WHO 2013, pp. 48, 67–8). Those working to stabilize the plant received the highest radiation doses, increasing their chances of suffering from cancer or leukaemia in the medium to long-term. The health ministry has subsequently identified that at least 63 workers were exposed to levels higher than had been recorded, and by the end of 2012 more than 160 workers had exceeded the maximum exposure limit of 100 millisieverts over five years (Myers 2013). This reflects the significant difficulties in accurately determining the amount of exposure during the period when radiation levels were at their highest. It is worth recalling that the vast majority of the workers were contractors, who often do not have accurate records of their exposure levels to begin with, leaving them more vulnerable in the future. One worker who had been employed by a subcontractor but was forced to leave, because he exceeded his exposure limit during the crisis, reflected that 'they used us and threw us away' (Negishi 2013).

The Dai-ichi workers could be described as 'double victims', as they suffered both from the tsunami and from being at the frontline of the nuclear crisis. Plant workers were not somehow immune from the horrible events of the tsunami. Many of their families, friends and colleagues were dead or missing. Their homes had been swept away or they had been forced to abandon them because of radiation, and like other surviving residents they had to live in difficult conditions at evacuation shelters or at the plant. Meanwhile, they were labouring to prevent the situation from further spiralling out of control. And when the reactors exploded, they feared they would die, but still they continued. One worker recalled that, 'the person who sent us back didn't give us any explanation. It felt like we were being sent on a death mission'. Another recounted that his manager used a military term – 'final battle orders' – when telling him to return (Birmingham and McNeill 2012, pp. 72–3). Indeed, this reflected the instructions of Prime Minister Kan, who ordered TEPCO not to abandon the plant, and made clear that the workers would have to lay their lives on the line to save the nation, if necessary. Having to deal with these incredibly difficult conditions not only impacted the physical health of the workers, it also had consequences for their mental well-being.

Due to the tireless efforts and sacrifices of the 'Fukushima Fifty', the worst-case scenario, in which Tokyo would have been evacuated (Fackler 2012), was averted. While some have rightly noted that there was a great deal of luck involved in avoiding this 'truly existential threat' (Funabashi and Kitazawa 2012, p. 19), it is necessary to stress the vital part played by those at the plant who did everything they could to stop the situation from deteriorating further. Much emphasis has been placed on it being a 'man-made disaster'; but it also worth remembering that there was a man-made solution.[3] Yet once the plant was stabilized, a new series of major challenges emerged: bringing it to cold shutdown and then decommissioning the gravely damaged reactors. This longer process has also had a serious impact on the human security of the workers.

Longer term insecurities

The decommissioning of Fukushima Dai-ichi is an unprecedented operation in terms of scale and difficulty. Conservative estimates suggest that it will take 30 to 40 years. To do this, TEPCO requires a large workforce, and one that will need to be constantly replenished as there is a high attrition rate due to poor conditions, and over the longer term also due to workers reaching their exposure limits. The decommissioning of Fukushima Dai-ichi is now being undertaken by a workforce of about 900 TEPCO employees and a further 5,000 contractors (McCurry 2013).

It is important to note that many of the vulnerabilities that defined the conditions of nuclear workers in Japan before the nuclear accident remain, and can be found in a particularly acute form at Dai-ichi. The contractors working at Dai-ichi suffer from the same problems as before: lower wages, limited rights and greater health risks. Due to the massive amounts of money TEPCO must pay in compensation, it has been forced to reduce salaries. According to a recent TEPCO survey, approximately 70 per cent of the workers made $9 per hour, less than what day labourers can earn elsewhere in the country (Saito et al. 2013). For those employed by subcontractors, their pay has been reduced through the various contractors first taking their cut. To give one example, a subcontractor had his monthly take-home pay reduced from $3,100 to $2,500 and his employer has stopped paying for his accommodation and meals. He rightly complained that, 'the plant remains stable, only because people like me are giving themselves up for the job. Asking them to accept poorer working conditions and driving them into a corner isn't fair at all' (NHK 2012). Even concerns about the low wages of contractors have not resulted in change: TEPCO announced in November 2013 that it would set aside another ¥10,000 per day for each worker, only for it to later clarify that this increase was 'for making efforts to improve workers' wages', thereby allowing contractors to take their cut and determine how much would actually reach the workers (*Mainichi* 2014c). Those who remain and accept these unfair conditions do so because they have few alternatives.

Not only does the subcontracting system allow vulnerable men to be economically exploited, it also increases the health risks they are exposed to. This is particularly consequential given that radiation levels remain much higher than in a usual working environment, and there are 'hot patches' around the plant where there are sudden spikes in radiation. There have been reported cases of subcontracted workers receiving poorer quality protective gear, radiation exposure records being incorrectly recorded and continued *yakuza* involvement (Slodkowski and Sato 2013). One former worker, Tetsuya Hayashi, took a job through a subcontractor for monitoring radiation exposure of workers, but on arrival at the plant he was assigned a different task in an area with a much higher amount of radiation. He explained that 'I felt cheated and entrapped. I had not agreed to any of this'. When he complained, he was fired (Slodkowski and Sato 2013). After these problematic hiring and work practices at Dai-ichi were exposed, Naomi Hirose, TEPCO's president, responded that changing the employment structure of relying on subcontractors would be difficult, concluding: 'it could *even* hurt the industry' (emphasis added, Yamaguchi 2012). This statement perfectly encapsulates the core problem from a human security perspective: after all the damage and suffering caused by TEPCO, their ultimate concern remains their narrow corporate interests, with little concern for the people they employ. The current subcontracting system may not hurt the industry, but it *does* hurt those vulnerable individuals that have few alternatives but to take these jobs.

TEPCO has not treated its own employees at Dai-ichi much better than contractors. Immediately after the disaster their salaries were cut by 20 per cent, as the company struggled to find money to begin paying for the damage caused. TEPCO is now seeking refunds of compensation payments made to its own employees that had been evacuated – 'double victims' that had to endure both the triple disaster and working at Dai-ichi as the disaster unfolded. At least 100 TEPCO employees have had their compensation payments stopped, and at least 15 of them have been requested to return money, totalling more than ¥100 million (*Mainichi* 2014a). In one case, an employee referred the matter to an alternative dispute resolution entity, which had been set up to deal with disaster compensation disputes. The non-binding ruling was in favour of the employee, but TEPCO has refused to follow it. According to workers at Dai-ichi, TEPCO's unsympathetic behaviour has significantly undermined their morale, and has led to people resigning out of anger and frustration (*Mainichi* 2014b).

TEPCO's poor treatment of its 'double victims' is an example of the way workers at Fukushima Dai-ichi must deal not only with significant physical health risks, but also considerable mental stress. Indeed, a recent WHO report observed that 'the psychological impact of the Fukushima accident may outweigh other health consequences' (WHO 2013, pp. 90–2). It also notes that the uncertain, ongoing nature of nuclear accidents leads to 'a high incidence of psychosomatic symptoms, psychological distress and psychiatric disorders', with 'parents with young children, pregnant women, children, elderly persons, emergency workers, people with pre-existing mental disorders, clean-up workers, evacuees' being most at risk (WHO 2013, p. 90). To date, this observation is proving accurate for the Fukushima nuclear accident. While the impact it has had on some of these groups – especially families with young children – has been well documented, there has been less attention given to the difficulties the workers of Dai-ichi have faced.

When considering the mental health of the Dai-ichi plant workers, there are a number of major stressors (Shigemura 2012). First, the job itself: they have to engage in physically exhausting work in very difficult conditions. Exposure to radiation is a worry that constantly plays on many of their minds. One subcontractor, describing how he feels, explains that 'I get stomach aches. I am constantly stressed. When I'm back in my room, all I can do is worry about the next day' (Myers 2013). Second, as noted, many also suffered from the tsunami – the death of loved ones, the loss of property, and being forced to evacuate. The grief and sense of displacement caused by these events does not easily disappear, and continues to affect them in the same way it does other survivors. Third, many have suffered from discrimination and abuse because they work for TEPCO. In these situations no distinction has been made between the company, its management and those that work at the plant: they are collectively seen as being responsible and have become targets of people's anger (Foster 2012; Nagata 2012). These experiences have weighed heavily on the workers, and according to one study it has been the biggest source of psychological trauma (Shigemura et al. 2012b, p. 668).

Given these difficult circumstances, it is unsurprising – though deeply troubling – that many workers have been suffering from mental health problems. According to one study almost half the Dai-ichi workers have shown symptoms of psychological distress, and post-traumatic stress response rates were very high:

> The workers showed a myriad of posttraumatic stress responses, including intrusive flashbacks, avoidance of their plant, hypervigilance toward aftershocks, fear of irradiation, and dissociative episodes. They showed immense grief and guilt owing to the deaths of their colleagues and loved ones.
>
> (Shigemura et al. 2012a, p. 1)

These observations correspond with the findings of research undertaken on Chernobyl clean-up workers, which concluded that they suffered from 'significantly higher rates of depressive disorder, anxiety disorder (without PTSD), and suicide ideation' (Loganovsky et al. 2008, pp. 484–6). While they did not suffer from noticeably greater instances of radiation-related cancer or disease, they were more prone to alcohol-related disease and more at risk of suicide (Rahu et al. 2013). There are already signs of the same problems appearing at Fukushima Dai-ichi, with reports of widespread alcohol abuse and depression amongst workers (McCurry 2013). These emerging mental health problems should be recognized and responded to before they worsen.

Human security for all

TEPCO is primarily responsible for the nuclear accident, but this does not mean that the well-being of workers at Fukushima Dai-ichi should be forgotten about or disregarded. The vast majority of people employed at Dai-ichi are subcontractors who must work in difficult conditions for limited payment, doing so out of pure economic necessity. Indeed, many of them are also victims of the tsunami, and they are equally deserving of the same support and sympathy offered to others. They perhaps may not be 'obvious' or 'good' victims in the way that mothers worrying about their children might be, but this is no reason to exclude them. It is worth recalling the broader conception of human security acknowledged in the 2010 Report by the UN Secretary-General, which entails 'freedom from fear, freedom from want and freedom to live in dignity for all' (UNSG 2010, p. 17). This reminds us that the workers should also be treated with the dignity and respect they deserve. They have had to deal with deeply traumatic experiences and continue to work in difficult, stressful circumstances. This reality should be acknowledged and properly responded to.

An important component of human security is prevention, which is particularly relevant to this case. The vulnerability of the workers at Dai-ichi is evident and steps should be taken to assist them. The difficulties of nuclear workers in Fukushima should not be downplayed or ignored simply because they are employed, directly or indirectly, by TEPCO. Discrimination against workers can be addressed and responded to (Shigemura 2012). This is particularly important given how significant an impact it appears to have on their mental well-being. There is also a need to build mental health support systems for these workers. This is a challenge, given the limited resources TEPCO has available and many specialists – including psychiatrists – not wanting to be associated with the company. Doing so will also entail overcoming a traditional culture that avoids directly talking about such problems.

Addressing the health of workers is important not only because their well-being matters, but also because they are doing vital work. Stress or trauma could result in human error, misconduct or even sabotage. There have been instances of contaminated water leaking due to mistakes made by workers, leading one former employee to warn that 'unless the working environment and working conditions improve, then another leak of contaminated water might occur (*Mainichi* 2013b). It will take decades before the plant is fully decommissioned and decontamination is completed. There are already reports that TEPCO has been having difficulty with maintaining the workforce required at Dai-ichi. It is only by adopting a less exploitative approach, one that is more concerned about the health and safety of its workers, that TEPCO will be able to successfully decommission the heavily damaged nuclear plant.

Conclusion

The most immediate threat posed by the failing Fukushima Dai-ichi plant has now passed, but in many ways the disaster is still ongoing. Former residents do not know when, if ever, they will be able to return. The Japanese government and the IAEA can only estimate when the plant will be fully decommissioned. The exact state of the reactors remains unknown because the radiation levels inside them are too high to allow proper investigation. While it seems the amount of radiation released into the air was limited, the exact impact the disaster will have on people's physical health remains unclear. Fierce debate continues about whether longer term exposure to low-level radiation is dangerous or not. Regardless, there are clear signs that the disaster has traumatized and affected the mental health of many Fukushima residents. As the rest of the country rebuilds, the workers at Fukushima Dai-ichi must continue to face the nuclear accident on a daily basis, as they struggle on in difficult conditions and with little reward.

In many cases the most vulnerable groups in society are hidden from our view, and this is certainly the case with the workers of Fukushima Dai-ichi. That they work – either directly or indirectly – for TEPCO does not mean the insecurities they are facing do not deserve our concern. Indeed, the fact that they are so central to stabilizing and decommissioning the plant means there are important pragmatic, as well as principled, reasons for supporting them. At the frontline of the disaster they have faced the highest doses of radiation, and thus are more vulnerable to future physical health problems, such as cancer or leukaemia. In addition, they have also had to deal with considerable stress, and many are deeply traumatized from their workplace experiences. A vast majority are employed as contractors, which means they receive less pay and are regularly exposed to greater health risks. While there is an obvious tendency to focus on radiation, in many ways it could be their exposure to stress that poses the greater human security threat.

It is important to note that many of the problems detailed here are not limited to the Dai-ichi power plant. The same phenomenon – vulnerable men being exploited through a complex system of subcontracting – is occurring in relation to decontamination projects in the surrounding areas. The huge amount of money allocated by the Ministry of Environment for decontamination contracts has resulted in a massive influx of companies seeking to profit, many of which have not been sufficiently vetted. The three major *yakuza* gangs are also alleged to be heavily involved (Saito and Slodkowski 2013, pp. 2–4). A shortage of workers has led subcontractors to actively prey on homeless men or those in debt. One homeless man explains: 'We're an easy target for recruiters … They say to us, are you looking for work? Are you hungry? And if we haven't eaten, they offer to find us a job' (Saito and Slodkowski 2013, p. 7). This offers another example of the complex ways that human security threats can be interconnected: the government's desire to have towns decontaminated as quickly as possible, so as to allow evacuees to return and restart their lives – which would contribute to improving their human security – results in the vulnerability of others being actively exploited by unscrupulous subcontractors.

At Chernobyl the people forced to do the most dangerous task of clearing debris close to the reactor were described as 'bio-robots', because real robots engineered for the task could not function as the radiation levels were too high (Rahu 2003, p. 296). The humanity of these workers was effectively denied as they were put at the service of the Soviet state. Ultimately the way workers have been used by TEPCO at Fukushima has not been that much different. As Naomi Hirose revealed in his complaint that labour reforms to ensure better treatment for workers 'could even hurt the industry', it is clear that TEPCO's corporate interests remain the primary concern, an approach that contributed to the disaster in the first place (NAIIC 2012). TEPCO and Japan must find a way to successfully decommission Fukushima Dai-ichi, but this should not come at the expense of the vulnerable group of people that must work there. Human security is meant to be a universal condition; and the insecurities of the nuclear workers should not be forgotten.

Notes

1 As of Nov. 2013, 1,605 people from Fukushima had died of indirect causes relating to the evacuation following the nuclear accident (*Mainichi* 2013a).

2 A particularly useful and detailed account of how the accident unfolded, which synthesizes all the major inquiries, is provided by Kushida (2012).

3 Gendered language is most appropriate here: TEPCO did not appoint its first female executive until 2013 (Torres 2013). Likewise, almost all the plant workers during the disaster were men, which corresponded to the general state of the nuclear industry. As such, gendered language is consciously used here and throughout the chapter.

References

Adelstein, J. 2012, 'How the Yakuza went nuclear', *Daily Telegraph*, 21 Feb., http://www.telegraph.co.uk/news/worldnews/asia/japan/japan-earthquakeand-tsunami-in/9084151/How-the-Yakuza-went-nuclear.html, viewed July 2012.

Aldrich, D. 2008, *Site Fights: Divisive Facilities and Civil Society in Japan and the West*, Ithaca, NY: Cornell University Press.

Birmingham, L. and McNeill, D. 2012, *Strong in the Rain*, New York: Palgrave.

Commission on Human Security 2003, *Human Security Now*, New York: CHS, http://isp.unu.edu/research/human-security/files/chs-security-may03.pdf, viewed Aug. 2013.

Fackler, M. 2012, 'Japan weighed evacuating Tokyo in nuclear crisis', *New York Times*, 27 Feb., http://www.nytimes.com/2012/02/28/world/asia/japan-considered-tokyo-evacuation-during-the-nuclear-crisis-report-says.html, viewed Dec. 2013.

Foster, M. 2012, 'Doctors: Japan nuclear plant workers face stigma', *Lexington Herald-Leader*, 5 Aug., http://www.kentucky.com/2012/08/05/2285637/doctors-japan-nuclear-plant-workers.html, viewed Aug. 2012.

Funabashi, Y. and Kitazawa, K. 2012, 'Fukushima in review: A complex disaster, a disastrous response', *Bulletin of the Atomic Scientists*, 68(2), pp. 1–13.

Hecht, G. 2012, 'Nuclear nomads: A look at the subcontracted heroes', *Bulletin of the Atomic Scientists*, http://thebulletin.org/nuclear-nomads-look-subcontracted-heroes, viewed Dec. 2013.

Jobin, P. 2011, 'Back to Fukushima', *HesaMag*, 4 (Autumn–Winter), pp. 35–9.

Kushida, K. 2012, *Japan's Fukushima Nuclear Disaster: Narrative, Analysis, and Recommendations*, Shorenstein APARC Working Papers, June, http://iis-db.stanford.edu/pubs/23762/Japans_Fukushima_Nuclear_Disaster.pdf, viewed Dec. 2013.

Loganovsky, K., Havenaar, J. M., Tintle, N. L., Guey, L. T., Kotov, R. and Bromet, E. J. 2008, 'The mental health of clean-up workers 18 years after the Chernobyl accident', *Psychological Medicine*, 38(4), pp. 481–8.

McCurry, J. 2013, 'Plummeting morale at Fukushima Daiichi as nuclear cleanup takes its toll', *Guardian*, 15 Oct., http://www.theguardian.com/environment/2013/oct/15/fukushima-nuclear-power-plant-cleanup, viewed Jan. 2014.

Mainichi. 2013a, 'Fukushima nuclear evacuation-related deaths surpass prefecture's quake, tsunami toll', 17 Dec., http://mainichi.jp/english/english/newsselect/news/20131217p2a00m0na010000c.html, viewed Dec. 2013.

Mainichi. 2013b, 'Ex-Fukushima worker warns of more leaks from hastily constructed tanks', 23 Dec., http://mainichi.jp/english/english/newsselect/news/20131223p2a00m0na012000c.html, viewed Jan. 2013.

Mainichi. 2014a, 'TEPCO seeks refunds of evacuation payments from employees, rejects ADR settlement', 4 Jan., http://mainichi.jp/english/english/newsselect/news/20140104p2a00m0na005000c.html, viewed Jan. 2014.

Mainichi. 2014b, 'TEPCO worker frustrated over company's treatment of employees', 4 Jan., http://mainichi.jp/english/english/newsselect/news/20140104p2a00m0na008000c.html, viewed Jan. 2014.

Mainichi. 2014c, 'TEPCO allows contractors to dip into "labor fund" increase', 5 Jan., http://mainichi.jp/english/english/newsselect/news/20140105p2a00m0na004000c.html, viewed Jan. 2014.

Myers, D. 2013, 'Cleaning up after Fukushima', *Digital Media News*, 10 March, http://dmnewsi.com/2013/03/10/reporters-notebook-cleaning-up-after-fukushima, viewed June 2013.

Nagata, K. 2012, 'Tepco workers face mental health crisis after cost cuts: Counsellor', *Japan Times*, 14 Aug., http://www.japantimes.co.jp/text/nn20120814a2.html, viewed Aug. 2012.

Negishi, T. 2013, 'Former TEPCO employee seeks donations for downtrodden Fukushima workers', *Asahi Shimbun*, 29 Dec., http://ajw.asahi.com/article/0311disaster/fukushima/AJ201312290006, viewed Jan. 2014.

NHK 2012, 'Fukushima plant workers are leaving the workforce', *NHK Today's Close-Up*, 5 Nov., http://www.nhk.or.jp/japan311/kuro-workers.html, viewed Dec. 2013.

NAIIC 2012, *The Official Report of the Nuclear Accident Independent Investigation Commission*, Tokyo: NAIIC, National Diet of Japan, http://www.nirs.org/fukushima/naiic_report.pdf, viewed Dec. 2013.

Onitsuka, H. 2011, 'Hooked on nuclear power: Japanese state–local relations and the vicious cycle of nuclear dependence', *Asia-Pacific Journal*, 10/2(2), http://www.japanfocus.org/-Hiroshi-Onitsuka/3676, viewed Jan. 2013.

Penney, M. 2011, 'Heroes or victims? The "Fukushima Fifty"', *Japan Focus*, 28 March, http://www.japanfocus.org/events/view/61, viewed Dec. 2013.

Pittmann, G. 2012, 'High distress seen in Japanese nuclear plant workers', *Reuters*, 14 Aug., http://www.reuters.com/article/2012/08/14/us-distress-nuclear-idUSBRE87D0X820120814, viewed June 2013.

Rahu, M. 2003, 'Health effects of the Chernobyl accident: Fears, rumours and the truth', *European Journal of Cancer*, 39(3), pp. 295–9.

Rahu, K., Auvinen, A., Hakulinen, T., Tekkel, M., Inskip, P., Bromet, E. J., Boice, J. D. Jr and Rahu, M. 2013, 'Chernobyl cleanup workers from Estonia: Follow-up for cancer incidence and mortality', *Journal of Radiological Protection*, 33(2), pp. 395–411.

Saito, M. and Slodkowski, A. 2013, 'Fukushima's murky clean-up', *Reuters*, 30 Dec., http://www.reuters.com/article/2013/12/30/us-fukushima-workers-idUSBRE9BT00520131230, viewed Jan. 2014.

Saito, M., Takenaka, K. and Topham, J. 2013, 'Japan's "long war" to shut down Fukushima', *Reuters*, 8 March, http://www.reuters.com/article/2013/03/08/us-japan-fukushima-idUSBRE92417Y20130308, accessed July 2013.

Shigemura, J. 2012, 'Why Fukushima nuclear plant workers are facing discrimination and stigma: Mental health consequences following the Fukushima Daiichi accident', presentation at 'Vulnerability and Empowerment: Rebuilding after 3/11' symposium, United Nations University, 30 Nov.

Shigemura, J., Tanigawa, T. and Nomura, S. 2012a, 'Launch of mental health support to the Fukushima Daiichi nuclear power plant workers', *American Journal of Psychiatry*, 169(8), p. 1.

Shigemura, J., Tanigawa, T., Saito, I. and Nomura, S. 2012b, 'Psychological distress in workers at the Fukushima nuclear power plants', *Journal of the American Medical Association*, 308(7), pp. 667–9.

Slodkowski, A. and Sato, M. 2013, 'Special report: Help wanted in Fukushima. Low pay, high risks and gangsters', *Reuters*, 25 Oct., http://www.reuters.com/article/2013/10/25/us-fukushima-workers-specialreport-idUSBRE99O04320131025, viewed Jan. 2014.

Tabuchi, H. 2011a, 'Japanese workers braved radiation for a temp job', *New York Times*, 9 April, http://www.nytimes.com/2011/04/10/world/asia/10workers.html, viewed Dec. 2013.

Tabuchi, H. 2011b, 'Economy sends Japanese to Fukushima for jobs', *New York Times*, 8 June, http://www.nytimes.com/2011/04/10/world/asia/10workers.html, viewed Dec. 2013.

Thébaud-Mony, A. 2011, *Nuclear Servitude: Subcontracting and Health in the French Civil Nuclear Industry*, Amityville, NY: Baywood Publishing Co.

Torres, I. 2013, 'TEPCO appoints first female executive amidst hard times', *Japan Daily Press*, 26 April, http://japandailypress.com/tepco-appoints-first-female-executive-amidst-hard-times-2627829, viewed Dec. 2013.

UNSG 2010, 'Human security, report of the secretary-general', UNSG A/64/701, 8 March, https://docs.unocha.org/sites/dms/HSU/Publications%20and%20Products/Reports%20of%20the%20Secretary%20General/A-64-701%20English.pdf, viewed Jan. 2014.

World Health Organization 2013, *Health Risk Assessment from the Nuclear Accident After the 2011 Great East Japan Earthquake and Tsunami, Based on a Preliminary Dose Estimation*, Geneva: WHO, http://apps.who.int/iris/bitstream/10665/78218/1/9789241505130_eng.pdf, viewed Dec. 2013.

Wingfield-Hayes, R. 2013, 'Why Japan's "Fukushima 50" remain unknown', *BBC News*, 3 Jan., http://www.bbc.co.uk/news/world-asia-20707753, viewed June 2013.

Yamaguchi, M. 2012, 'Japan utility admits murky hiring at nuclear plant', *Associated Press*, 10 Dec., http://bigstory.ap.org/article/japan-utility-admits-murky-hiring-nuclear-plant, viewed Dec. 2013.

5 Human security as a military security leftover, or as part of the human condition?

Paul James

Sometimes the obvious is obscured by detailed attention. Sometimes the accretion of constant use buries the obvious in the glitter and dirt of taken-for-granted past assumptions. This is very much the case with our understanding of human security. The adjective 'human' in 'human security' is treated as completely obvious and no more than a gentle qualifier of the term 'security'. Since the first breakthrough documents in the field of human security, 'the human' has been substantially taken for granted and never defined. Certainly the question of what is human security has evoked much discussion and debate. However, despite this direct attention, it has become inexplicably rare for the obvious semantic core of the concept to be examined. To do so would be to begin to interrogate the meaning of 'the human' in ways that would challenge dominant paradigms.

Rather than getting buried in debates over the narrow version of security – namely, freedom from fear – versus the broad version of freedom both from fear and want, this chapter takes a couple of steps into the task of redefining human security by treating security and risk as part of the contingent and negotiated condition of human living. In these terms, the liberal notion of 'freedom from', for all its importance, is exposed as a limited reduction of the richness of positive human security. Emphasizing what we want to escape 'from' is to remain with the classical notion of *negative freedom*. Adding to this the concept of 'freedom to', i.e. *positive freedom*, where people begin to debate the complex meaning of what it is to live with the tensions of both security and risk, is to take one step forward conceptually.

Taking freedom out of its currently privileged *singular* place at the centre of the discussion is to take a further step forward into a more complex sense of human flourishing. The alternative negotiated themes of human security that need to be brought into contention include the relationships between freedom and obligation, between security and risk, between needs and limits and between authority and participation, amongst many others. In relation to the security/risk dialectic, for example, there can never be total security from the risk of natural disaster. To project such a scenario of 'zero risk' as TEPCO and the Japanese Nuclear and Industry Safety Agency did for the Fukushima reactors prior to their meltdown is political myth-making (Suzuki 2011). To attempt to achieve such a practical impossibility is a folly and would be to begin to annul the terms of a second tension between needs and limits. On the other hand, treating security/

risk as simply a cost-benefit ratio is to rationalize human needs as if they can be turned into an economic calculus of life and death, suffering and compensation.

Human security in its fullest sense, including relative security from extreme natural events, needs to be understood as a variable condition of the human – what the 1994 *Human Development Report* called the 'universalism of life claims' or 'human life for itself' (UNDP 1994). In other words, if this thing called 'human security' has the concept of 'the human' embedded at the heart of it, then let us address the question of the human condition directly. Thus understood, human security would no longer be the vague amorphous add-on to harder edged areas of security such as military security or state security. Neither would it be reduced to a few complementary factors such as food security, economic security and personal security, linked by the notion of 'freedom from want'. It would become the centre of all discussions of security. It would become the basis of thinking about all security – including military security and disaster management.

Critics suggest that human security approaches are damned because the concept lacks a precise and generally agreed definition. Certainly, confusing debates surround the concept. However, by the end of the present chapter I hope that it will become obvious that this is an unhelpful critique. What we can say rather is that the concept of 'human security' has joined the ranks of what have been called 'essentially contested concepts'. Old concepts such as 'democracy' and 'justice', and new concepts such as 'globalization' and 'human rights', have become deeply contested because they are foundational to contemporary political and normative debates. The task is then a twofold one: first, to examine the strengths and weaknesses of existing discourses and definitions; and secondly to reframe the approach to human security by taking forward its strengths and delimiting its weaknesses. This will be done against a backdrop discussing how we might better respond to crises, from slow global crises such as climate change to immediate and more localized crises such as the Fukushima disaster in Japan following the 2011 Tōhoku earthquake and tsunami.

The limits of the paradigm shift

The paradigm shift that made possible the first conceptions of human security, like all paradigm shifts, was based on an emergent and important ideological movement rather than a single pronouncement. Across the course of the twentieth century, documents such as the constitution of the International Labour Organization, the United Nations Charter and the Universal Declaration of Human Rights linked peace and security to broader social justice issues (Therien 2012). Within this broad change in sensibility, present in at least one lineage of the sustainable development and social justice movement, it is possible to point to key moments and documents that mark the development and consolidation of the new conception of human security. Three foundational documents can be singled out: the 1994 *Human Development Report*, the 2003 report, *Human Security Now*, and the 2010 UN Secretary-General's 2010 report on human security (UNDP 1994; Commission on Human Security 2003; UNSG 2010).

These documents are relatively consistent in their thinking, and as a body of work they have many strengths. They make a fundamental break in expressing human security in terms of an explicit shift of emphasis from the security of *the state* to the security of *people*. There is, for the most part, a nice ambiguity in the documents about the concept of 'people', allowing it to encompass both peoples as communities and people as individuals. This is actually unusual in the literature. To be blunt, when secondary writings sometimes assume that individuals are the basis of human security, the authors of those writings tend to be exposing their grounding assumptions in political liberalism, with liberalism's emphasis on individuals rather than communities of people. By comparison, the original documents were, and are, more open. Moreover, in some parts of the originating documents, the writing is positively generalizing in defining human security in terms of 'fundamental freedoms' *and* 'fulfilment'. This is broader than just freedom from want and freedom from fear, as human security is often evocatively – and reductively – summarized (including in some other parts of these very same documents). The term 'fulfilment' begins to evoke the sense of positive engagement that takes us to a fuller debate about the human condition.

Paradoxically, like many paradigm shifts, the founding documents leave much intact from previously inadequate approaches. The emperor they address has one set of clothes – liberal modern – and while the new paradigm has tailored those clothes with gentle humanistic language, the broader social imaginary that colours the paradigm is still reductive, rationalizing and modern. In these terms, the problems with many existing discussions of human security can be summarized into a number of key points.

1. The relationship between human security and state security is left as one of weak complementarity between two security domains of equal standing

Despite the path-breaking work done on the concept, proponents of human security remain imprisoned by existing dominant discourses. Instead of taking the concept of 'human' as relating to the human condition and treating state security (and therefore military security) as being encompassed by human security, they tend to treat human security and military security as separate but equal domains. In the context of contemporary international relations this seems positive. It seems to counter contemporary dominant trends. Over the last few decades, the dominant approach has been to consolidate state security and military security as having such common-sense prominence that the more recent concern with human security has struggled to find even a complementary place alongside those violence-related regimes. Thus, assertions of parity or complementarity seem to be radical in their challenge to the dominant paradigm.

For example, the report *Human Security Now* suggests that, 'Human security *complements* state security, enhances human rights and strengthens human development' (CHS 2003, p. 2, emphasis added). Or again, 'Human security and state security are *mutually reinforcing and dependent* on each other' (p. 6, emphasis added). Read quickly, these statements make sense. But in each of the proposals, *state* security has somehow been retained at the same categorical level as the other much broader considerations of human development and human rights. It is as if the state-based security (and its military apparatus, itself a subordinate category of state practices) has a separate life from the human. When military security is subsequently brought back into a complementary relationship to human-wide categories such as rights, development and security it becomes their equal because of a self-confirming process. We need military security because our militarily insecurity is intensifying. It is easy to see why this is the case given the extraordinary emphasis over the decade or so on the Global War on Terror and Homeland Security, both often set out with capital letters. The obvious paradox is that massively increasing global spending on military security has not made the world more secure – perhaps the opposite.

In this context it is worth asking the question why disaster security, which should have an equal footing with military security, is a secondary regime of human activity. States continually reproduce the conditions for engaging in military activity, while, for the most part, states delegate disaster engagement to a relatively marginal team who work on response-preparedness plans. Japan is often said to be one of the most disaster-prepared states in the world. It has comprehensive plans for responding to disasters and an integrated network of administrators from the Central Disaster Management Council, chaired by the Prime Minister, linked down to the prefectures and the municipalities. In 2001 a Ministry of State for Disaster Management was established to coordinate disaster management policies and disaster responses. And yet the Japanese Self-Defence Force had an annual budget of over US$55 billion in 2011–12, while the budget for disaster risk reduction in 2010–11 prior to 3/11 was US$11billion, about 1 per cent of the total national account expenditure, down from 6 to 8 per cent of the national budget in the 1960s. The ratio in most other countries is stretched even further. The United States currently has an annual defence budget of US$689 trillion while its Federal Emergency Management Agency gets US$6.8 billion, an absurd 10,000:1 ratio (Global Firepower 2013; Tomata and Sagara 2012).

In this context the concept of 'peace' has a parallel problem. Once the problem is pointed out, it only takes a moment to reflect on how peace has come to be defined without positive qualities. However, finding or even formulating a positive definition is very difficult. In all our dictionaries, peace is defined negatively as the absence of war. *War* has become the master signifier and *peace* no longer has a publicly accepted definition of its own. Similarly, human security has come to be defined by all those security areas that are not encompassed directly by state or military security. Human security, like peace, is thus what is *left over* after the war-machine has been given full definitional, practical and financial pre-eminence.

There is a simple and overriding point that is missed here. The 'human' as a category encompasses, *by definition*, considerations of governance, the state and the military – and all the domains of social life beyond. Therefore *human* security should include (and subsume) state and military security, rather than sit alongside those other considerations as yet another subcategory of security. When children play 'category' games they implicitly understand such issues of ordering. In the simple terms of such games the categories of 'human' and 'human security' are to 'military security' or 'food security' as the categories of 'animal' and 'mammal' are to 'lion' or 'wildebeest'. If this is so by the logic of language, why is it not the case by convention and application?

The fact that it is not the case has major implications for how we respond to disasters. Disasters, including the slow disaster of climate change, are no less consequential in prospect and outcome than military threats. They are no less broad, and therefore in the 'category tree' of security issues the threat of disasters and military should sit at the same level. Currently this is not even close to being the case. If taken seriously, this rethinking would have significant consequences for convention and application. It would mean that threat assessment and resourcing of preparation for disasters, as well as responding after disasters, would become as important as preparing for and going to war.

It is possible, of course, that with the increasing incidence of storm surges, inundation and extreme weather events brought upon the earth by intensifying climate change that the emphasis on disaster management will increase, but this is certainly not the case at the moment. Japan is located on the Ring of Fire, the major earthquake line that runs around the Pacific, and is in an area of weather volatility. It is subject to multiple natural forces: earthquakes, including the Great Hanshin-Awaji Earthquake that killed over 6,000 people; tsunamis, including the Tōhoku tsunami that killed around 15,000 people; typhoons that have killed thousands of people; flooding, intensified by its steep fast-flowing rivers; and forest fires, based on large forested areas being close to densely populated areas. The irony is that after more than half a century during which Japan, against world trends, qualified the global emphasis on military security – and after one of its biggest disasters – it is now returning to a war-footing emphasis.

2. The category of the 'social' (i.e. of the human) is treated as one domain among many

Despite the active and engaged attempts by the proponents of a human security paradigm to get beyond the overwhelming emphasis in this world on military and economic issues, they tend to bring an emphasis on such issues back into contention by using the very same frameworks which are used by those who actively elevate economic and military concerns to the pinnacle of human interest. For example, the 2010 UN Secretary-General's report argues that

> the guarantee of national security *no longer* lies in military power alone. Essential to addressing security threats are also healthy political, *social*, environmental, economic, military and cultural systems that reduce the likelihood of conflicts, help overcome the obstacles to development and promote human freedoms for all.
>
> (UNSG 2010, emphasis added)

On the face of it the statement sounds strong. However, there are a number of problems that relate to our first point above. As the seemingly innocent phrase 'no longer' suggests, the statement builds on a background argument that because of increasing global interconnection it is 'no longer' the case that national security can be built on military power alone. The problem with this suggestion is that it is a distraction. It has never been the case that military power brought security by itself. National security has always relied for its resilience and sustainability upon much more than military security, including geography, natural boundaries, community resilience, cultural interchange across state borders, diplomacy, a robust economy and so on. The problem is redoubled by the list that follows: 'political, *social*, environmental, economic, military and cultural systems'. Here the 'social' and the 'economic' and the 'military' sit alongside each other at the same categorical level, subordinating the social within a laundry list of other things including military systems.

This example would be no great issue if it were not part of an overwhelming pattern. The usual mantra-like version of this kind of listing is the triple bottom-line categorization of 'economic, environmental and social' sustainability. Extra things are sometimes added to this triplet according to the orientation of the speaker or writer – military power or technology being the usual additions. In the corporate-oriented approach of the triple bottom line, the social – i.e. the way in which humans live and relate to each other and to nature – is effectively a secondary collation of things that do not fit elsewhere in the other categories. The economic is given an independent status: one that is ideologically assumed rather than analytically argued. In the most problematic versions, the economic is elevated to the master category and defined in terms that assume the dominance of a singular, historically specific, economic configuration: modern globalizing capitalism.

Concurrently the environment comes to be treated as an externality or background feature. It is an externality that tends not to have the human dimension built into its definition, as if we are not part of a complex ecology of species and systems. This has profound consequences for thinking about human security and disasters (Magee et al. 2013). Disasters are thus treated reductively as simply the consequence of extreme environmental events. More appropriately, disasters need to be understood as the consequence of the intersection of *social* questions, such as lack of organizational preparedness, community engagement, political adaptability, technical risk assessment and infrastructure appropriateness – *and* an extreme environmental event. Presented from the other way round, we can say that when extreme environmental events occur in places either where humans do not live or where the population and their infrastructure are well prepared then the events are not usually called 'disasters'.

3. Notwithstanding all the rhetoric to the contrary, human security tends to be reduced to the question of material security. Most of the human security literature has nothing more than high-sounding phrases to contribute to the question of existential security

Security and insecurity are integral to the broad ontological condition of being human. Security can never be an absolute, and insecurity takes many forms and levels of intensity. For a person or community experiencing deep ontological insecurity it does not matter whether or not they have enough food to stave off hunger or shelter over their heads to keep the rain out. For a person or community fearful that their cultural existence is subtly under threat, it does not necessarily console them that their material needs are all being met. And to complicate things further by a different example, for a God-fearing people, fear is part of the condition of their existence. It is fear that they fear losing. Despite this complexity, most of the literature on human security tends to focus on the material conditions of embodied security. Even the discussions of freedom from fear tend to concentrate on the leftovers from the centring of military and state security. Freedom from fear becomes freedom from fear of military action either by another state or one's own state. This criticism depends upon an analytical distinction between two interrelated dimensions of human security, material and existential. Material security concerns embodied needs – nutritious food, clean water, shelter, bodily integrity and so on. Existential security concerns sense and sensibility. More than just a psychology of safety, existential security is founded upon a sense of place, a sense of engagement, a sense of meaningful relations and so on. Together, material and existential security add up to ontological security in general.

The effects of the 2011 tsunami on children have been discussed in terms of psychoanalytic notions such as PTSD – post-traumatic stress disorder, a concept used as a category of analysis in Japan since the 1990s – and attention has been drawn to school essays by children with titles such as the 'Tsunami was black and smelly' (written by a second grader), 'The rumble in the earth' (by a fifth grader) and 'Listlessness, nausea, and a funny feeling' (by a sixth grader) (Hayashi and Tomita 2013). These are expressions of *personal* existential insecurity directly related to a disaster, but the process of existential insecurity includes this level and goes deeper. The effects of the Japanese bubble crisis, the global financial crisis, precarious employment and the triple disaster around the 2011 tsunami have all been having profound impacts on the ontological security of people in everyday life. When combined with changing social relations in contemporary Japan, the impact is compounded. These existential effects thus go beyond the materiality of economic precarity or even personal fear of a further earthquake, a tsunami or radiation poisoning. A slow social crisis of meaning is now evident, particularly for the younger generations living in cities. Japan Inc., brand-power, consumption affluence and the national state no longer provide the sense of existential security that they once did, even if only superficially, under conditions of capitalist growth. In Anne Allison's words:

> Hopelessness and futurelessness are buzzwords of the times. So is the phrase 'relationless society' (*muenshakai*): a descriptor of a country where the stitching of connectedness between people is fraying at the seams. Being alone – literally, psychically, socially – is the new human condition for Japan/ese in the 21st century. In an age when people are expected to be 'responsibly independent' (*jiko sekinin*), one-third of the population lives alone and 32,000 die 'lonely deaths' (at home all alone) – the same number as those who commit suicide – every year.
> (Allison 2009)

It is no accident that one of the most common concepts used since 2011 in Japanese disaster-relief campaigns is *kizuna*, meaning ties or bonds. Certain words have begun to be used by the state to name what is ideologically needed to respond to such crises, without the technocratic state any longer being able practically to effect such bonds or even engender political trust from the general citizenry (Dudden 2012). Emperor Akihito might be called upon to give speeches about pulling together as a nation. But for all the people who loved his speech, as many others dismissed it as an expression of the state ideological apparatus.

4. The definition of human security is shackled to the modern concept of 'freedom' and thus treated reductively as an expression of modern desires for choice

Human Security Now is an exemplary liberal document. According to the document, freedom is not only central to human fulfilment. It is also what makes it possible:

> The Commission on Human Security's definition of human security: to protect the vital core of all human lives in ways that enhance human freedoms and human fulfillment. *Human security means protecting fundamental freedoms – freedoms that are the essence of life.* It means protecting people from critical (severe) and pervasive (widespread) threats and situations. It means using processes that build on people's strengths and aspirations. It means creating political, social, environmental, economic, military and cultural systems [NB there is that list again] that together give people the building blocks of survival, livelihood and dignity. *The vital core of life is a set of elementary rights and freedoms people enjoy.*
> (CHS 2003, p. 4, emphasis added)

What should be noted is that here freedom is made the basis of human security. Having made a categorical modern claim, the report then relativizes that claim – *except in relation to freedom.* Freedom remains the universalizing bedrock. The passage continues: 'What people consider to be "vital" – what they consider to be "of the essence of life" and "crucially important" – varies across individuals and societies. That is why any concept of human security must be dynamic' (2003, p. 4). Absolutely. However, despite this apparent openness, the various dialectical tensions that have been argued about across human history, particularly since the eighteenth century – grounding tensions between needs (expressed as freedoms to choose) and limits, autonomy and authority, liberty and equality – are not here considered relevant.

5. The relationship to human rights tends to reduce human security to the liberal notion of 'freedom from' ('freedom from basic insecurities') as one among the many modern human rights

It is only a very short step from suggesting that freedom is the basis of human security to specifying the kind of freedom – freedom *from* basic insecurities. In

Human Security Now, Amartya Sen writes that 'freedom from' is the core concept of human security:

> Commitments underlying human rights take the form of demanding that certain basic freedoms of human beings be respected, aided and enhanced. The basically normative nature of the concept of human rights leaves open the question of which particular freedoms are crucial enough to count as human rights that society should acknowledge, safeguard and promote. This is where human security can make a significant contribution by identifying the *importance of freedom from basic insecurities* – new and old. The descriptive richness of the considerations that make security so important in human lives can, thus, join hands with the force of ethical claims that the recognition of certain freedoms as human rights provides.
>
> (2003, p. 9, emphasis added)

Human rights are normatively or ethically generated, but here they remain overgeneralized. Here human security just *is* – a descriptively rich and powerful claim that can 'join hands' with an ethical imperative.

6. Because of the liberal shackling of human security to modern notions of 'freedom', there is little or no understanding of insecurities that cut across ontological boundaries

The questions being posed here relate to the limits of the conventional liberal definition of human security, understood in terms of freedom from inequalities, deprivations and material denial of resources. This definition takes us a considerable way towards understanding the conditions of *modern* insecurity in the world today. However, there are profoundly good reasons to insist that security also exists across other layers of ontological practice and meaning – not just the modern. It is certainly the case that the material insecurity that fixes the attention of conventional approaches is ontologically foundational. But it is also worth considering the existential aspects of human insecurity from a less one-dimensional perspective.

Here one step in our argument is to claim that insecurity is caught in the intensifying tensions of intersecting social formations as processes of globalization have thrown together practices and sensibilities associated with different life-worlds. This might be called a clash of ontologies. The reference here to Samuel Huntington's *Clash of Civilizations* (1998) is as critique rather than as an extension of his approach (Senghaas 2002). The clash referred to is not that between ill-defined 'civilizations', nor is it a dubious variation on the 'ethnic differences cause violence' argument. Rather, the clash refers to tensions that arise at the intersection of different ways of being in the world – customary, traditional, modern and postmodern ways of living temporally, spatially and corporeally. These ways of living are not understood as separate domains but as layers of meaning and practice, criss-crossing the world, and often held together at the same time in the bodies and minds of people who find themselves moving across different life-worlds. For example, modern Western proponents of a certain kind of remembering responded to the terrorist disaster of the Bali bombing by putting up stone memorials. They did not understand that to fix the event as evil and to codify those killed by the bombs as victims was, at least for some, to confirm the pain for all time as an immutable presence. By ontological contrast, as Jeff Lewis and Belinda Lewis argue:

> Hindu-Vedic Balinese cosmology, known as the *rwa bhineda,* delicately suspends the forces of good and evil in an infinite and irresolvable dialectical combat ... Within this cosmology, however, evil cannot be subjugated, redeemed or eradicated from the material or spiritual world, since it is always and inevitably the predicate of good. To this extent, even the *bhuta kala* demons are not considered intrinsically evil, nor are their spiritual nemeses considered entirely wholesome. The *rwa bhineda* conceives of good and evil in terms of a mutual identification, whereby the faithful will seek merely to minimize the harm that evil may inflict on the living and the dead.
>
> (Lewis and Lewis 2009)

In Balinese mythology, stone monuments do not allow that delicate dialectical combat to work its way across time.

One crucial aspect of the clash of ontologies is that globalizing *modern* practices and meanings – including modern politics of security management – are caught in tension with the continuity of *traditionalism* and *customary* relations, which themselves have been stretched globally in new ways. Traditional religions provide a good example of this process. They are at once carried across the globe by modern processes of organization and communication, and they increasingly give rise to a modern politics of ethnic-cultural identification, but they remain grounded in traditional truths and authority claims that can be anathema to modern rights and freedoms.

Tensions between ontologies do not have to be negative; creatively handled such tensions can provide a positive ground to handling disasters, particularly where mortality becomes a significant issue. Claims to the eternal meaning of subjectivity given by God, Allah or the teachings of Shinto and the Buddha, and subjugation to positive fear of the transcendental, may be in direct tension with modern practices of abstract citizenry and individual freedom, but these alternative approaches to meaning can be brought into productive tension traversed through processes of cultural translation, including through appropriate monuments and public rituals. This means that the pretensions of liberal or postmodern cosmopolitanisms to sweep away these kinds of tensions – for example, by arguing for a world shackled to liberal freedoms (including freedom from want) – have qualified traction in practice. To the extent that cosmopolitanism remains unreflexive about such tensions, it cannot provide a grounded answer to the insecurities of our time.

In this area the *Human Development Report* is sensitively written, but it has nowhere to go conceptually, and therefore it remains limited practically. What we get is a liberal balancing act:

> Most people derive security from their membership in a group – a family, a community, an organization, a racial or ethnic group that can provide a cultural identity and a reassuring set of values. Such groups also offer practical support. The extended family system, for example, offers protection to its weaker members, and many tribal societies work on the principle that heads of households are entitled to enough land to support their family – so land is distributed accordingly. But traditional communities can also perpetuate oppressive practices: employing bonded labour and slaves and treating women particularly harshly. In Africa, hundreds of thousands of girls suffer genital mutilation each year because of the traditional practice of female circumcision.
>
> (1994: 31)

For all the balanced concern expressed here, there is no understanding about why customary practices become distorted under the globalizing pressure of modern practices and meanings (Geschiere 1999). The increase in genital mutilation discussed in the report is certainly a basic question of human security, but equating it with customary circumcision misses out on how we might best respond to the phenomenon in broad social terms.

The relevance to a country such as Japan, with its intertwining relations of custom, tradition and modern rationalities, is profound. Arguably, the more traditionally oriented people of Tōhoku, particularly people living in less urbanized areas, were better able to cope with the deprivation and the strictures of crisis-management authority after the 2011 tsunami than those living in the megacities. It has even been noted in the general press that, if an earthquake of the kind predicted for Tokyo actually happened, the ensuing social upheaval would be chaotic by comparison. The point that I am making is that people need more than the mechanics of material disaster relief in order to cope with such crises as deaths. That additional support does not necessarily have to come from traditional cosmologies that respond actively to mortality, but telling someone that they have PTSD and offering a session of counselling is not a sufficient replacement. Describing the Great Kanto Earthquake of 1923 that killed over 100,000 people, Gennifer Wiesenfeld writes: 'Bereaved families and the religious community, however made sure that the memories of their loved ones were not swept away in modernity's erasures; they foregrounded the primary sacred function of the memorial to honor and appease the spirits of the dead' (Wiesenfeld 2012).

7. Because of the modern emphasis on freedom from insecurity, there is a paradoxical giving with one hand and taking away with another

The possibilities that are 'given' by modern analytical (including liberal) approaches are extraordinarily important. In relation to natural disasters, modern systems – when they are communicated well – give us the fragile security of knowing in advance what is about to hit us, and the objective security of vacating a vulnerable area before a natural event occurs. Modern technological platforms measure and forecast the possibilities of disasters and the parameters of what they might entail. By sensor deployment, for example, we know when a tsunami or earthquake is coming, and, by using mathematical algorithms and modelling against known variables, we know its probable consequences. It is called 'probabilistic risk assessment'. We can thus prepare.

The 'taking away' is nevertheless deeply problematic. First, probabilistic risk assessment usually fails to take into account non-linear, indirect and compounding or other risk issues. It gives the impression of safety, defended by scientific verity, as if probabilities are actualities. Combined with the false assurances of self-interested bodies such as the nuclear power industry, it takes away from the fact that risk is ever-present. Secondly and more importantly, modern techno-science and management subsumes every understanding of nature and its effects within an 'imminent frame' that tends to reduce disaster response to a technical task. It tends to present a technocratic process shorn of the meaning of cultural difference and seeking to overcome nature. This is not to suggest the 'end of culture' or the 'end of nature' (to evoke the phrase used by Bill McKibben 1989). Rather, in the sense used by Charles Taylor (2007), it is to suggest that when modern scientific secular understandings frame disaster relief something is lost. To give a secular example, in contemporary disaster management, science and the politics of safety tend to override prior senses of identity, culture and livelihood. In the case of Sri Lanka, after the 2004 tsunami no-build buffer zones (200 metres from the sea in the north and east; 100 metres elsewhere) were legislated, overriding the long history of some communities for whom being on the beach was central to their way of life.

Natural crises such as floods, tsunamis and earthquakes were once framed by traditional cosmologies (religions) that gave existential sustenance in the face of ongoing fear. The biblical story of the flood relates to a covenant with God, as does the Mesopotamian story of Gilgamesh. Japan's stories about its earthquakes are related to the way in which its islands sit upon the catfish Namazu, flopping about, constrained only by a huge rock controlled by the god Kashima. These cosmologies now tend to be relegated to the edges of modern social life, demarcated in ritual observance or referred to in quaint reminders such as the use by the Japan meteorological society of a stylized catfish in its Earthquake Early Warning logo. This means that, in relation to natural disasters, while we have the technological capacity to predict and marginally contain the impact of tsunamis, we are offered little more than nominally enhanced *material security*. To the extent that the modern notion of 'freedom from' offers negative freedom, we find ourselves increasingly alone in the world, dependent on what are often now perceived as unreliable technological systems and fallible experts. This gives rise to a different kind of fear.

Towards an alternative framing of human security

The generality of the human (the social) can, for the purpose of operationalizing an alternative, be analytically broken down into domains and subdomains of the social, beginning with the primary domains of economics, ecology, politics and culture. In this 'circles of social life' approach, these domains are all understood as social and linked to the natural/social tension over themes such as the relation between needs and limits. Rather than the environment being reduced to an externality of the economic, it becomes the general ground on which we all walk. In turn, ecology becomes defined as the intersection between the social and the natural, and is not treated either as something outside of the social or as a modern rationalized way of understanding what it calls a natural system. This definition then emphasizes the role of human agency in the effects and consequences of all ecological events, including 'natural disasters'.

This analytical schema is still modern in its conception, but it provides a more holistic basis for responding to any comprehensive disaster – in realization of present social relations and social-natural tensions, in anticipation of the risks of future natural extreme events, or in the aftermath of an extreme event that has disastrous consequences.

Step 1

We need to treat 'the human' or 'the social' as the foundational category of life – which means that human security is relevant across the domains of the social. Each of the four social domains presented below is divided into seven subdomains. Disaster thinking would need to work through all of the subdomains.

Economics

The economic domain is defined as the practices and meanings associated with the production, use and management of resources, where the concept of 'resources' is used in the broadest sense of that word.

1 Production and Resourcing
2 Exchange and Transfer
3 Accounting and Regulation
4 Consumption and Use
5 Labour and Welfare
6 Technology and Infrastructure
7 Wealth and Distribution

Ecology

The ecological domain is defined as the practices and meanings that occur across the intersection between the social and the natural realms, focusing on the important dimension of human engagement with and within nature, but also including the built environment.

1 Materials and Energy
2 Water and Air

3 Flora and Fauna
4 Habitat and Settlements
5 Built Form and Transport
6 Embodiment and Sustenance
7 Emission and Waste

Politics

The political is defined as the practices and meanings associated with basic issues of social power, such as organization, authorization, legitimation and regulation. The parameters of this area extend beyond the conventional sense of politics to include not only issues of public and private governance but more broadly social relations in general.

1 Organization and Governance
2 Law and Justice
3 Communication and Critique
4 Representation and Negotiation
5 Security and Accord
6 Dialogue and Reconciliation
7 Ethics and Accountability

Culture

The cultural domain is defined as the practices, discourses, and material expressions, which, over time, express continuities and discontinuities of social meaning. (See Akiko Fukushima's chapter in this volume on the central role of culture.)

1 Identity and Engagement
2 Recreation and Creativity
3 Memory and Projection
4 Belief and Meaning
5 Gender and Generations
6 Enquiry and Learning
7 Wellbeing and Health

In such a schema, it can be readily seen that military security has a first-order relevance to only one subdomain of one domain – namely in the political subdomain of 'security and accord'. Of course, it quickly should be added that military security has a broader (second-order) relevance to all of the domains of politics, economy, ecology and culture, and their various subdomains, but then so does food security, for example. In the contemporary globalizing world of nation-states and translocalized violence, military security (or food security or housing provision, etc.) provides one necessary (but not sufficient) material condition of human security that connects all other dimensions of the social.

Working through all of these subdomains of the social in relation to disaster management would mean that planning would need to become more integrated and involve people in thinking about what it means to face disaster long before it actually happens – economically, ecologically, politically and culturally. Community resilience and volunteerism could not be reduced to capital that can be drawn upon when a disaster strikes, like money out of the bank. Instead of having an earthquake disaster plan, a fire plan and a climate change adaptation plan for events that are projected into the future, themes of memory and projection, belief and meaning, would be evoked as issues to be discussed now. Planning for ecological outcomes would be related together in a systematic way, and planning would become more than setting up *response* plans for some future possible action. Activities conducted continually (rather than simply in anticipation of a disaster) would bring cultural questions about security and risk into the centre of contention, including dialogue about our engagement with nature, and how to represent it creatively, symbolically.

Step 2

We need to incorporate sensitivity about the ontologically variable way in which each of these social themes is practised and understood. This includes a sensitivity to the ontological variability of the notion of 'risk' itself (and not in the all-encompassing sense that Ulrich Beck (1992) describes the contemporary modern condition as a 'risk society'). All societies have been risk societies, monitoring and responding to risk. Ours does so by modern processes of mediated monitoring, electronic codification, abstract mapping and probability assessment. The customary cultures of the Trobriand Islands do so by being on the beach and responding to the immanent-transcendental movements of the waves, the fish and the spirits.

Step 3

We need to redefine 'security' and 'insecurity' so these concepts no longer rest on modern liberal absolutes: 'freedom from want; freedom from fear'. Rather than being absolute, human security is grounded in processes of social sustainability and social resilience. Rather than being absolute, security is relational. In these terms, insecurity – both material and existential insecurity – can sometimes be a good thing, as can uncertainty, struggle and adversity, so long as they do not attack the foundation of what it means to be positively alive and related to others in enriching ways. This becomes particularly relevant in the present when the classical modern high-romantic sensibility of the sublime (which includes the dialectic of the fear and love of nature) has been largely supplanted by the late-modern techno-scientific sensibility that nature can be either controlled, or at least substantially moderated by geo-engineering precautions and responses. The phrase, 'prevention, response, remediation' has become a slogan without the proponents really working through what these terms mean beyond being technical steps in a giant Gantt chart. In the context of climate change it becomes increasingly important that risk assessment does not get reduced to carbon accounting and meteorological metrics. A complex sense of *insecurity* would be a much more sane response to the current situation than the current mainstream confidence that, in our ingenuity and innovation, security-based assessments will eventually lead us to respond adequately to the comprehensive challenge that faces us on this planet.

Conclusion

This brings us, in conclusion, to a new definition of human security. Summarizing the previous discussion, human security can be defined as one of the foundational conditions of being human, including both (1) the sustainable protection and provision of the material conditions for meeting the embodied needs of people, and (2) the protection of the variable existential conditions for maintaining a dignified life. Within this definition it then makes sense that the core focus of human-security endeavours should be on the most vulnerable. It makes sense that risk management should be most responsive to immediate events or processes that have both an extensive and intensive impact in producing material and existential vulnerabilities of people in general or a category of persons across a particular locale. Within this definition we can thus analytically distinguish two kinds of human security: material security and existential security. Going deeper, we can also distinguish between two modalities of human security. *Negative* human security concerns the process of overcoming the violation of human rights. *Positive* human security concerns the process of sustaining the variable and contingent conditions of vibrant human life across *all* the domains of social life – economic, ecological, political and cultural.

References

Allison, A. 2009, ‘Ordinary refugees: Social precarity and soul in twenty-first century Japan’, *Anthropological Quarterly*, 85(2), p. 345–70.

Beck, U. 1992, *Risk Society*, London: Sage Publications.

Commission on Human Security. 2003, *Human Security Now*, New York: CHS, http://isp.unu.edu/research/human-security/files/chs-security-may03.pdf, viewed Aug. 2013.

Dudden, A. 2012, ‘The ongoing disaster’, *Journal of Asian Studies*, 71(2), pp. 345–59.

Geschiere, P. 1999, ‘Globalization and the power of indeterminate meaning: Witchcraft and spirit cults in Africa and East Asia’, in B. Meyer and P. Geschiere (eds), *Globalization and Identity*, Oxford: Blackwell, pp. 211–238.

Global Firepower 2013, ‘Countries ranked by military strength 2013’, www.globalfirepower.com/countries-listing.asp, viewed Dec. 2013.

Hayashi, K. and Tomita, K. 2013, ‘Lessons learned from the Great East Japan Earthquake: Impact on child and adolescent health’, *Asia-Pacific Journal of Public Health*, 24(4), pp. 681–8.

Huntington, S. 1998, *The Clash of Civilizations and the Remaking of World Order*, London: Simon & Schuster.

Lewis, J. and Lewis, B. 2009, 'Recovery: Taming the *Rwa Bhineda* after the Bali bombings', in D. Grenfell and P. James (eds), *Rethinking Insecurity, War and Violence: Beyond Savage Globalization?* London: Routledge, pp. 194–207.

McKibben, B. 1989, *The End of Nature*, New York: Anchor Books.

Magee, L., Scerri, A., James, P., Thom, J.A., Padgham, L., Hickmott, S., Deng, H. & Cahill, F. 2013, 'Reframing sustainability reporting: Towards an engaged approach', *Environment, Development and Sustainability*, 15(1), pp. 225–43.

Senghaas, D. 2002, *The Clash within Civilizations: Coming to Terms with Cultural Conflicts*, London: Routledge.

Suzuki, T. 2011, 'Deconstructing the zero-risk mindset: The lessons and future responsibilities for a post-Fukushima nuclear Japan', *Bulletin of the Atomic Scientists*, 67(5), pp. 9–18.

Taylor, C. 2007, *A Secular Age*, Cambridge, MA: Harvard University Press.

Therien, J-P. 2012, 'Human security: The making of a UN ideology', *Global Society*, 26(2), pp. 191–213.

Tomata, M. and Sagara, J. 2012, *Knowledge Note 6-1, Measuring the Cost-Effectiveness of Various DRM Measures*, Washington, DC: World Bank.

UNDP 1994, *Human Development Report 1994*, New York: Oxford University Press.

UNSG 2010, 'Human security, report of the secretary-general', UNSG A/64/701, 8 March, http://hdr.undp.org/en/media/SG_Human_Security_Report_12_04_10.pdf, viewed Aug. 2013.

Wiesenfeld, G. 2012, *Imagining Disaster: Tokyo and the Visual Culture of Japan's Great Earthquake of 1923*, Berkeley, CA: University of California Press.

6 Human security and life recovery

Lessons from the 1995 Kobe Earthquake and the 2011 triple disaster

Mayumi Sakamoto

Introduction

Natural disasters occur suddenly and change lives. Natural hazards such as earthquakes or tsunamis are difficult to predict, but once they happen they leave tremendous damage. People suddenly lose houses, property, family, friends, health and jobs. Displacement caused by the disaster forces survivors to live in totally different environments: evacuation shelters, temporary housing and then new permanent housing. The lives of survivors are often disrupted many years after the initial disaster. Until survivors move into new permanent housing, their lives are generally insecure.

The Great Hanshin-Awaji Earthquake, which hit Kobe City and the surrounding area in Hyogo Prefecture, Japan, on 17 January 1995, with a magnitude of 7.3 on the Richter scale, was the first mega-disaster to hit a city in Japan during the post-Second World War period. It left widespread damage in its wake: 6,434 persons died, more than 43,792 persons were injured and 639,686 buildings collapsed. A number of new problems were identified during and after the recovery, which are relevant from a human security perspective.

The first was *Saigai Kanren-shi* ('disaster-related death'). The dead numbered 6,434; within this number, 5,502 people (85 per cent) died as a result of the direct impact caused by the earthquake. However, another 940 people (15 per cent) died several days/weeks/months/years after the earthquake because of the indirect effects of the natural disaster, such as the deterioration of their health due to illness or injury, psychological stress caused by the disaster, infections in evacuation shelters or respiratory problems caused by debris removal. After the Hanshin-Awaji Earthquake, these post-disaster deaths were termed 'disaster-related deaths'. Victims of disaster-related death were recognized as disaster deaths and received the same benefits from the compensation system as other disaster victims.

The second was *Kodoku-shi* ('isolated death'). Huge temporary housing complexes were constructed in the suburbs of Kobe. Those who had lived in evacuation shelters wanted to move to temporary housing as soon as possible. Temporary housing residents were selected by lottery using a priority system, with first priority given to the most vulnerable, such as the elderly or people with disabilities. As a result of this policy, vulnerable people were concentrated in several temporary housing complexes. In this situation, an overarching concern for vulnerable people actually resulted in an increase in their vulnerability, because of flaws in policy design. Elderly people who lived alone in temporary housing died without their deaths being noticed by neighbors. In order to prevent isolated death, a life support system known as the Life Support Advisor (LSA) was created. LSAs visited temporary housing to check the health condition of residents.

The third was *Saigai Shogai* ('disaster-related disabilities'), referring to people who became disabled due to injury or illness caused by the disaster. In the Hanshin-Awaji Earthquake, there were people trapped for hours or days under collapsed buildings; many of these people suffered fractures, spinal cord injuries, bruises and so on. Some of them could not get sufficient medical care because of physical damage to hospitals, and a lack of medical staff or medicine. The lack of health and social support for those injured in the disaster made their living conditions worse. However, their problems remained obscured for years.

The experience of the Hanshin-Awaji Earthquake revealed that, for human security in natural disasters, both pre-disaster and post-disaster preventive efforts to secure human lives are required. Before the disaster it is important to prevent physical damage and protect people from the direct impact of natural hazards. However, once a disaster occurs it is important to shift policies to prevent further health damage during the recovery process, and to support the life recovery of survivors.

Life recovery was a new idea realized in the recovery process after the Hanshin-Awaji Earthquake. Life recovery is a people-centered concept and entails people returning to a feeling of normalcy in their lives. Life recovery in the case of Hanshin-Awaji Earthquake survivors encompassed a range of factors such as *housing, social ties, community, health, preparedness, economy and governance*. Life recovery requires an interaction among these factors. A recovery policy which focuses only on physical recovery, such as buildings, roads and water reconstruction, but not on people, is liable to fail. Recovery should integrate people, society and safety. This policy approach is reflective of human security's concern for relational qualities, which entails connecting individuals into the fabric of their societies. Human security is generated not simply through material disaster relief, but it is also generated through a range of social aspects that influence a person's general well-being and psychological disposition.

The Great East Japan Earthquake occurred on 11 March 2011, and was the biggest earthquake/tsunami disaster in Japan after the Hanshin-Awaji Earthquake. The disaster-stricken areas had suffered repeated earthquakes and tsunamis in the past. Therefore, disaster mitigation and preparedness efforts such as large-scale seawalls for tsunami protection, tsunami hazard maps, evacuation routes, evacuation drills and disaster education had been prepared and were used. It was the most disaster-prepared area in Japan. However, the Great East Japan Earthquake differed from the Hanshin-Awaji Earthquake. It was a multi-location/multi-hazard disaster. The unexpected scale of the tsunami destroyed seawalls, embankments, bridges, roads, buildings and houses. About 600 km of the Pacific coast of east Japan was totally destroyed. Were the lessons from the Hanshin-

Awaji Earthquake applied in the Great East Japan Earthquake? If so, why was it difficult to reduce the damage?

This chapter focuses on human security in the disaster response and recovery phases, through the experience of these two different mega-disasters in Japan. First it reviews the recovery process after the 1995 Great Hanshin-Awaji Earthquake, a mega-earthquake which struck the City of Kobe, focusing on the post-disaster lives of people, to clarify the lessons obtained. Next it reviews the recovery process after the 2011 Great East Japan Earthquake, the multi-location/multi-hazard disaster. It then discusses how lessons learned as a result of the Hanshin-Awaji Earthquake were applied in the recovery process, what problems remain and what is required to secure human lives.

The Great Hanshin-Awaji Earthquake

Sudden and unexpected large tremors struck the City of Kobe on 17 January 1995, at 5:46am. Kobe is a beautiful port city located in central Japan, with a population of 1.6 million people. Almost 16 seconds of tremors overwhelmed the city. Buildings and houses collapsed; roads, transportation systems and water supplies were destroyed, and the city turned into debris. What made the situation worse were fires caused by broken power lines that burned down dense residential areas of wooden houses.

Figure 6.1 indicates the causes of death related to the earthquake. Under the weight of heavy buildings, 4,978 persons (77 per cent of the total dead) were thought to have died immediately after the tremor through suffocation or crushing. Wooden houses, such as two-story wooden row houses established from around

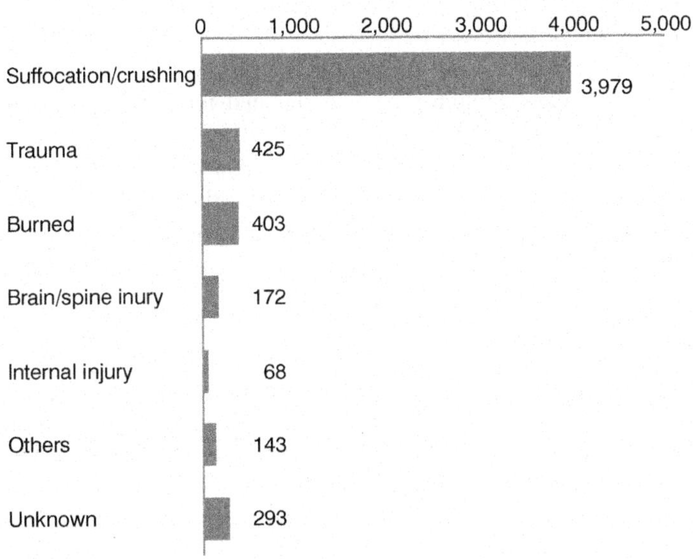

Figure 6.1 Causes of death in the 1995 Great Hanshin-Awaji Earthquake

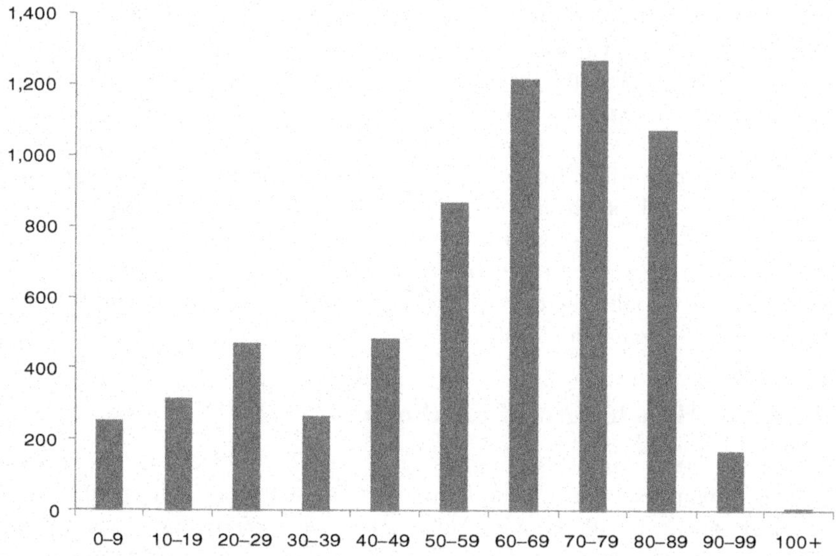

Figure 6.2 Breakdown of death by age in the 1995 Great Hanshin-Awaji Earthquake

1950 to 1960, were severely damaged. A large number of the dead were elderly people (Figure 6.2). It is assumed that those elderly lived on the first floors of old houses in the stricken areas. Since the rent was cheaper in these old wooden houses, vulnerable people such as the elderly and students suffered disproportionately.

Aftershocks continued, and people who had lost their houses or worried that their houses would collapse during the aftershocks took refuge in evacuation shelters. According to the local disaster management plan, public buildings such as gymnasiums, school classrooms and community centers were to be used as evacuation shelters since these buildings were designated for public purposes and constructed to earthquake-resistant standards. The number of evacuated people was greatest on 23 January 1995. A total of 1,158 evacuation shelters were established, and 316,678 persons were evacuated to these shelters (Figure 6.3). The school gymnasium floors were full of bedding, and many disaster survivors stayed there together.

Life in evacuation centers

The living environment in the evacuation shelters was said to be awful. Since public buildings such as school gymnasiums were not planned to be used as evacuation shelters, toilet, shower and kitchen facilities were insufficient. Many people were gathered in closed spaces with no privacy. No specific attention was paid to gender issues, people with disabilities or mothers with infants. Women faced the problem of where to change clothes, and mothers where to nurse babies. There were not enough bathrooms. Temporary toilets were set up; however, most

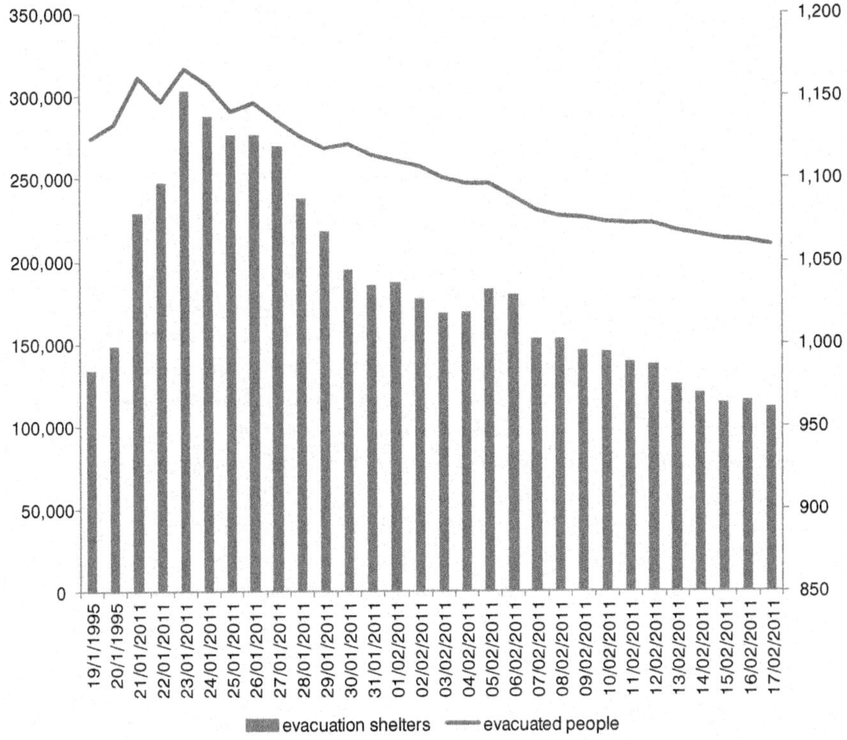

Figure 6.3 Evacuation timings in Hyogo Prefecture (19 January 1995–17 February 1995)

of them were traditional style, with steps at the entrance, which made it difficult for elderly people to use them. As a result, some people refrained from drinking water, and some elderly persons were obliged to use diapers.

It was obvious that such bad living conditions affected people's health. In order to minimize the damage to those who were living in evacuation shelters, voluntary medical rehabilitation teams, composed of medical doctors, nurses, physiotherapists, occupational therapists and others, visited the shelters. They evaluated and treated people living in shelters. These visits were conducted from 17 January to 31 March 1995 to evacuation shelters in the Hanshin-Awaji region (Mizuno 1995). In the evacuation shelters, there were at least 1,003 persons who faced functional problems in daily life and needed rehabilitation. Among the 1,003 persons who received support from the team, 146 persons (15 per cent) were healthy before the disaster but required medical rehabilitation because of injuries caused by the disaster; the other 201 (20 per cent) were healthy before the disaster and were not injured, but needed rehabilitation for physical or psychological conditions after the disaster. In Kobe City, of 465 persons who received treatment, 235 persons required follow-up treatment; the conditions of 76 patients (32.3 per cent) improved, 128 patients (54.5 per cent) remained the same and 10 patients

(4.3 per cent) experienced deterioration in their conditions. Patients were offered training for functional recovery, daily activity instruction and aids such as canes were provided. In addition, an improvement in nutrition, psychological support and the living environment of the shelters made a significant difference. Where deteriorations in the conditions of patients occurred, this was because of diseases such as influenza and pneumonia, and because of falls and broken bones and bruises due to the gymnasiums' slippery floors.

Infectious diseases like influenza spread easily in the evacuation shelters. The stress caused by changes in the living environment caused people's health to deteriorate. According to a study by Ueda, a doctor at Kobe Kyoudou Hospital, 104 persons were sent to his hospital within four weeks of the earthquake (Ueda 1996). Of these, 18 were hospitalized, 11 of whom were from evacuation shelters. Their average age was 77.7. The main diseases were pneumonia and bronchitis (45 patients). Among the hospitalized patients, 18 patients died because of illnesses contracted during their time in evacuation centers, and these were recognized as disaster-related deaths.

There is a compensation system for natural disaster victims and for those who lose family members in a natural disaster. This system provides a maximum of ¥5 million (about US$50,000). After the Hanshin-Awaji Earthquake, the system also came to be applied to victims of disaster-related death. In order to identify that the natural disaster was the reason for death, special committees comprised of medical doctors, lawyers and government officials were organized. As of January 1996, the number of disaster-related deaths in Kobe City was 615 and 69.6 per cent of those who died were over 60 years old. In order to prevent disaster-related deaths, it is important to improve post-disaster living environments, especially in evacuation shelters. A human security approach to natural disasters should recognize the special needs of vulnerable people in these settings.

Temporary housing

In order to move survivors out of evacuation shelters, the Kobe municipal government began construction of temporary housing three days after the earthquake. However, since Kobe was a city developed on a narrow plain between the Rokko Mountains and Seto Inland Sea, it was very difficult to find adequate space for temporary housing in the city center. Land for temporary housing was developed on the outskirts of Kobe, far from where survivors had lived. Of the temporary housing 29,178 units were constructed in Kobe City and 3,168 houses outside the city (Kobe City 2000). People who had lived in evacuation shelters wanted to move to temporary housing as soon as possible. It was difficult to provide temporary housing for all of them right away; the municipal government decided to provide housing in five phases (Table 6.1).

In order to avoid confusion over selection for temporary housing, the government established a priority framework (Table 6.2). Households with vulnerable persons, such as those with only elderly individuals (more than 60 years old), persons with serious disabilities, or mothers and children, were the first to apply.

Table 6.1 Applications for temporary housing

Phase	*Subscription period*	*No of houses*	*No of applications*	*Houses provided*
1	27/1/1995–2/2/1995	2,702	59,449	2,340
2	28/2/1995–7/3/1995	12,802	63,367	8,458
3	7/4/1995–11/4/1995	6,740	25,794	4,748
4	10/5/1995–14/5/1995	4,095	16,683	3,625
5	1/7/1995–6/7/1995	10,028	13,989	8,656

Source: Compiled from data provided by Kobe City

Table 6.2 Priority criteria for temporary housing following the 1995 Great Hanshin-Awaji Earthquake

Priority	*Criteria*
1	Households with only elderly people (over 60 years)
	Households with people with serious disability
	Households with mother and child/children under 18 years
2	Households with elderly people (over 66 years)
	Households with infant (under 3 years)
	Households with pregnant woman
	Households with more than 3 children (under 18 years)
3	Households with person with injury/illness
	Households with person with vulnerable health condition
4	Other households

Source: Compiled from data provided by Kobe City

The first phase of application for temporary housing was from 27 January 1995 to 2 February 1995, during which time 59,449 applications were received, and 2,340 households were selected. Houses were provided gradually and completed in August 1995. All evacuation shelters were closed on 21 August 1995. After people moved to temporary housing, two unexpected problems arose. First, a society of mainly elderly people had formed. The Kobe municipal government conducted a study in February 2006 that revealed that among all residents in temporary housing, 41.7 per cent were more than 65 years old, and 70.6 per cent had low incomes (lower than ¥3 million). In other words, the temporary housing community was potentially very vulnerable. Second, pre-disaster communities were separated, and it became difficult to rebuild new communities. Since houses were constructed on the outskirts of Kobe City, far from where survivors had previously lived, and residents were selected by lottery, strangers came to live next to each other. It was difficult for some elderly people to establish new relationships in a new community. This became a cause of isolated death.

Generally, two kinds of cases can be considered isolated death (Shimokawa 2011). First, persons living alone without any communication with neighbors or families died alone without anyone realizing. Second, persons living alone but in contact with neighbors or families died alone due to a sudden deterioration in their health. The cases observed in Kobe were mostly the former, in which elderly persons living alone in temporary housing died alone as chronic illnesses became more serious and they did not receive proper health assistance. Their neighbors did not realize this because they did not have any communication with them. In Kobe City, 132 persons were confirmed as victims of the isolated death phenomenon in 2001 (Kobe City 2010).

In order to prevent isolated death, two support systems were established. First there was the Life Support Advisor (LSA) system, to support people living in temporary housing. LSAs visited temporary housing to check on the health of residents and to facilitate procedures to improve their environment. Second was the revitalization of the community through the creation of social networks. Volunteers or NGOs contributed to such networking by organizing community meetings, events or small businesses to support people living in temporary housing. These community-building support systems were also continued with regard to permanent housing. However, isolated deaths also occurred in permanent housing, which once again repeated the same fragmentation of community.

The experience of the Hanshin-Awaji Earthquake revealed that policies which focus only on vulnerable people but not on community security can fail. People do not live alone. They belong to some kind of community. In each community, people are linked through different kinds of social networks, whether family-based, neighborhood-based, school-based or job-based. These networks are invisible and embedded in society but function in many ways, and it was such networks that provided support for vulnerable people in daily life. When people did not see a neighbor for days, someone would go to check on his/her health condition. In the case of Kobe, social networks in pre-disaster communities were suddenly and artificially cut off during the recovery process because of policies that did not focus on the provision of community security. This relates closely to the original understanding of human security developed in the 1994 UNDP report, which emphasized that it 'is concerned with how people live and breathe in a society'. A human security approach to natural disasters should exhibit an awareness of community security, of the importance of social ties and relationships during the recovery process.

Disaster-related disabilities

It has been assumed that there were people who suffered injuries and illnesses in the Hanshin-Awaji Earthquake and became disabled for years. Newspapers have reported the problems of such people with disaster-related disabilities; however, the exact number of victims and their living situations is not clear (Sakamoto 2011). In order to support people who became disabled because of injury or illness caused by natural disaster, the Japanese government established a 'natural disaster-related disability compensation system' in 1982. In addition to the usual social welfare system for people with disabilities, the system provides financial support of ¥2.5 million (about US$25,000) maximum to people who became seriously disabled, such as losing sight or mobility in both arms, due to natural disasters. The system's benefits are limited to seriously disabled persons whose opportunities for social participation are limited, and for whom it would be difficult to recover from their disability through medical treatment or rehabilitation.

In the Hanshin-Awaji Earthquake, 43,792 persons were injured, and within this number 10,683 were seriously injured. However, as of 2010, only 61 persons, those identified by Hyogo Prefecture as having disaster-related disabilities, had received compensation. The compensation system applied only to the most disabled persons. Some people with disaster-related disabilities might have been difficult to identify because of personal information protection. Journalists repeatedly reported problems of victims who lacked government social support, referring to the 'forgotten problem'. It was because of this that Hyogo Prefecture and Kobe City governments decided to conduct a detailed study on people who became disabled because of the disaster. The study was conducted in 2010–11. By checking application records and interviewing survivors, the government found that 328 persons (131 male, 197 female) were newly identified as disabled because of injury or illness caused by the disaster. Within this number, 192 persons (58.5 per cent) were over 60 years old; 121 persons had already passed away.

The study revealed several new facts. First was that those with disaster-related disabilities faced more than just health problems. Of those disabled, 63 per cent were injured because of house destruction, and 69 people (73.7 per cent) had lost their houses. Also 19 people (22 per cent) had lost one or more members of their family who lived with them, and 16 people (19 per cent) had family members who were injured. Twenty-four people (29 per cent) had lost their jobs, and 18 people (21 per cent) changed their working status, for example, from full time to part time. Second, the study noted the lack of social support for people with disabilities. Throughout the interviews, many complained that there was no social support after they got injured. The study revealed that people who were recognized as having a disaster-related disability within a year of the earthquake were just 44 per cent of those eventually identified. In order to receive social support, people must get medical certification confirming that their disability is permanent; they receive social welfare benefits based on that medical certification. This meant that people could not receive any support until their disability was certified. Third, the study discovered a lack of accessibility. Physical disability affected not only mobility but also social participation. This is an example of the way different kinds of vulnerability can interact negatively. In order to rebuild their lives after the disaster and receive support, people had to follow several procedures including producing documentation. People had to get certificates for damaged housing and disaster victim certificates in order to apply for temporary housing, change of schools and so forth. Those procedures had to be carried out in different departments of the municipal government office, and it was difficult for seriously injured persons to get the required information and conduct the various procedures without help. One survivor said, 'how can a person like me who cannot walk apply for support?'

Life recovery

The biggest lesson learned from the recovery process following the Hanshin-Awaji Earthquake was the importance of *life recovery*. Roads, houses and buildings could be reconstructed almost as scheduled, but it was difficult to recover people's lives. Human security is about more than purely material needs. The Kobe City Recovery Plan had around 1,000 recovery projects, which were roughly classified according to six elements.

Figure 6.4 shows the structure of the recovery projects (Kobe City 2000). Infrastructure reconstruction refers to reconstruction of public facilities which sustain civil life, such as water supplies, roads and railways. Infrastructure is vital, and it took almost two years to reconstruct. Housing reconstruction and urban planning were important to secure displaced people's lives. It took almost five years to move people into permanent housing. Economic revitalization and small and medium enterprise (SME) reconstruction secured survivors' income generation; this took almost 10 years to recover. All these elements were considered prerequisites to help survivors rebuild their lives.

Life recovery means that people return to a sense of normalcy, an idea that closely matches the way human security is conceived in the 1994 UNDP report. Our daily lives are composed of various elements, and there are several critical factors that influence life recovery as a whole. In 1999, a series of grassroots stakeholder assessment workshops were conducted by the Kobe City Research Committee

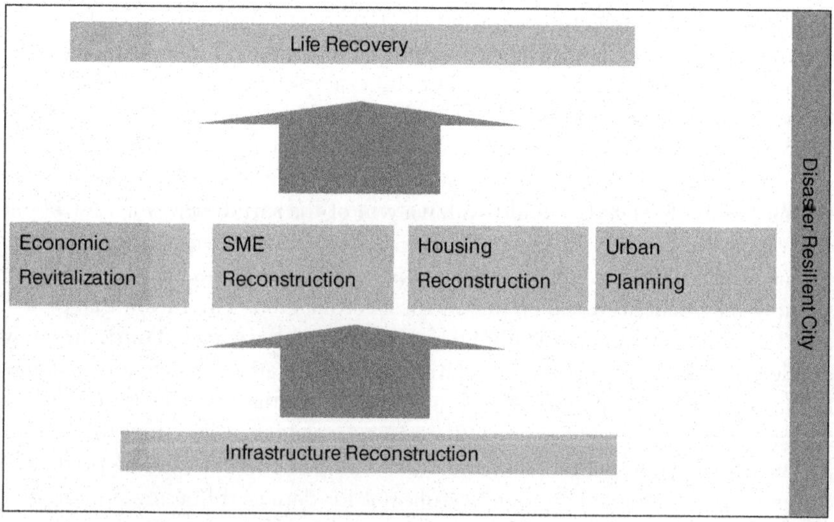

Figure 6.4 The basic structure of Kobe City recovery projects

on Disaster Recovery, with the participation of not only disaster survivors, but academics and local government officials as well, in order to identify life recovery determinants (Tatsuki 2007). Overall, 269 people participated in a total of 12 workshops, and 1,623 opinion cards were collected in the process. The cards were carefully studied and classified according to the following seven critical elements: housing, social ties, community rebuilding, physical and psychological health, preparedness, economic and financial situation and relation to government.

Such factors were examined during workshops in 2001, 2003 and 2005. The most critical factor in the 1999 workshop was 'housing', followed by 'community rebuilding' and 'preparedness'. However, 'housing' declined in importance in the 2003 workshop, with 'social ties' becoming the top priority, followed by 'community rebuilding' and 'preparedness'. As a result of this important comprehensive multi-year survey, it became clear that the most important factor for life recovery over the long term was 'social ties', articulated in opinions such as the following: 'I have been helped', 'Understanding and empathy through ties', 'New relationships emerged', as well as 'Self-governance and solidarity are keys to developing closer ties' (Tatsuki 2007).

Using quantitative data, Tamura also verified the results. Regarding social ties, the people who felt self-governed and expressed solidarity with others felt that they had recovered more fully (Tamura et al. 2001). In the short term, issues such as housing and income, as well as physical and psychological health management, support life recovery and directly alleviate the damage caused by the disaster. Additionally, social ties facilitate event evaluation, i.e. they help to establish the meaning of the experience. In turn, event evaluation is influenced by social ties built through community rebuilding, empowerment and/or opportunity enrichment, both due to encountering significant people who are capable of providing assistance. The 17 years of recovery experience following the Hanshin-Awaji Earthquake revealed that people's lives were insecure throughout the recovery process. People's lives exist within communities, and each community's individuals are linked through social networks. Recovery policy that separates existing communities' social networks is liable to fail. It is important to focus not just on vulnerable people but also society as a whole, and to put people's lives at the center of the entire recovery policy. This closely matches the 'people-centered' approach of human security.

The Great East Japan Earthquake: a multi-location, multi-hazard disaster

The Great East Japan Earthquake occurred on 11 March 2011, causing widespread damage in east Japan, and leaving 18,812 persons dead or missing as of 4 July 2012. The disaster-stricken areas also suffered repeated earthquakes and tsunamis. Therefore, disaster mitigation and preparedness efforts such as the large-scale seawalls for tsunami, tsunami hazard maps, evacuation routes, evacuation drills and disaster education had been prepared and were used. However, the scale of the tsunami exceeded damage estimation predictions. The huge tsunami destroyed damage-mitigating facilities such as seawalls or four-story buildings constructed with reinforced concrete for tsunami evacuation, threatening the security of people who had evacuated to places that were considered to be safe.

Compared to the Hanshin-Awaji Earthquake, the disaster response by the national and local governments was quick. The local government set up Disaster Management Centers immediately after the first tremor was observed. They announced tsunami evacuation immediately. The national government established the National Disaster Management Headquarters 30 minutes after the first tremor. Two hours later, all the representatives from the ministries headed to disaster-affected Miyagi Prefecture by helicopter, and they established the Onsite Disaster Management Headquarters at the Miyagi Prefectural government office on 12 March at 6:00am.

The present Japanese disaster response plan is based on the Hanshin-Awaji Earthquake experience. However, disaster response for the Great East Japan Earthquake was difficult because the damage pattern was complex and different from that of the Hanshin-Awaji Earthquake. It was a multi-location, multi-hazard disaster. In the case of the Hanshin-Awaji Earthquake, the damage was concentrated in some cities but was not area-wide. In the case of the Great East Japan Earthquake, on the other hand, the damage was spread across wide areas and long-term support for disaster survivors was required. To make the situation worse, several unexpected problems occurred.

First, many local governments lost their offices or staff and did not have enough capacity to support survivors. In Japan, disaster response is primarily the duty of municipal government. In case a municipal government cannot manage the situation, it requests support from the prefectural government. However, a situation in which several local governments lose capability had not been discussed before the disaster. Second was the accident that occurred at the Fukushima nuclear power plant. It was damaged by the tsunami, and a series of explosions occurred. The national government had to concentrate their attention on preventing a nuclear meltdown and evacuating people from the area. The accident became an obstacle to mobilizing resources to disaster-stricken areas. Some companies refused to transport resources because they would have to go through Fukushima Prefecture. Local NGOs hesitated to support disaster survivors from Fukushima. International rescue teams who came to Japan left not only Fukushima Prefecture but also Japan as they were afraid of radiation damage. Finally, the lack of oil and damage to critical infrastructure was severe. The earthquake and tsunami damaged the oil storage and processing facilities that were constructed along the Pacific coast. In the disaster-stricken areas, roads were totally covered with rubble and the harbors filled with debris. Thus, no vehicles or ships could enter the area until the rubble was cleared.

Evacuation life

The living situation of the Great East Japan Earthquake survivors was critical, especially in the first three weeks after the tsunami. Evacuation shelters were set

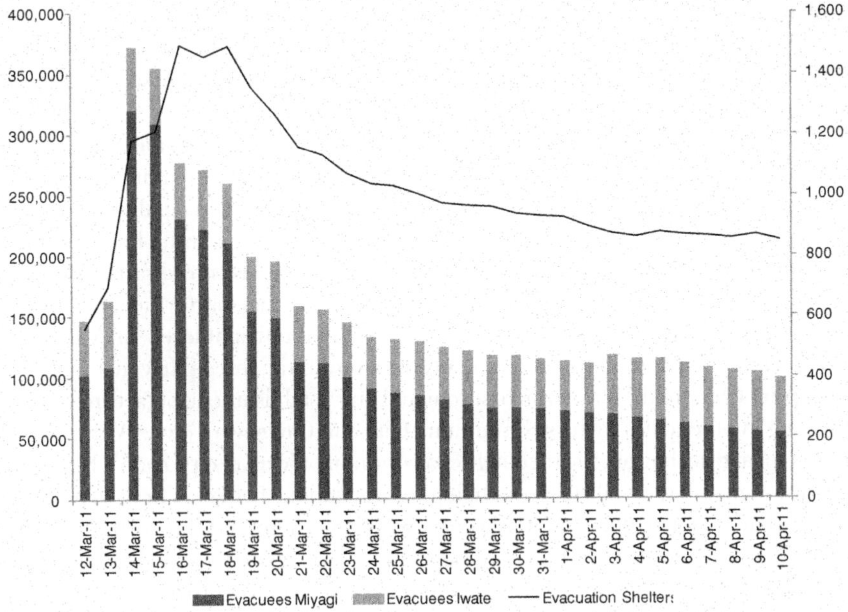

Figure 6.5 Evacuation Timings in Miyagi Prefecture and Iwate Prefecture (12 March 2011–10 April 2011)

up in municipalities along 600 km of Pacific coast. Because of the devastated roads, transportation damage and lack of fuel, it was difficult to transfer food or daily commodities to all survivors. Local governments had lost offices and staff because of the tsunami, and it was difficult to provide support. By the end of March, although weeks had passed since the earthquake, there were still more than 100,000 people in evacuation shelters who could not get sufficient food or daily commodities (Figure 6.5).

In this situation, several efforts were made to support survivors. First, the government and NGOs/NPOs collaborated to support survivors (Sakamoto 2011). Evacuees, especially those who were in shelters of between 1,500–2,000 people, had not eaten a hot meal since the disaster had occurred. Only two *onigiri* (rice balls), twice per day, were distributed to them. The Japan Self-Defense Force tried to provide food, water and showers to these evacuees; however, they did not have enough resources to provide hot meals to everybody on a daily basis. In some evacuation shelters, residents did not want to receive hot meals unless there was sufficient food for everyone. Several NGOs/NPOs wanted to provide hot meals; however, they did not know which evacuation shelters to serve. Thus, it became evident that, if all institutions were coordinated, better support could be provided. This led to the initiation of the *Hisaisha Shien Yonsha Renraku Kaigi* ('four-party meetings in support of disaster survivors', hereafter referred to as 'four-party meetings'). 'Four-party meetings' involved the national government's onsite headquarters, the Self-Defense Force, the Miyagi Prefectural government and representatives from various NGOs/NPOs. The four-party meeting was established as a result of an initiative from the government's onsite headquarters. There had never been such collaboration before; however, the strong will to support the survivors made the meeting a reality.

In order to make the coordination practical at the local level, the *Sansha Kaigi* ('three-party meeting') was organized in Ishinomaki City, Kesennuma City and Onagawa Town, attended by the Self-Defense Forces, the municipal government and NGO/NPO representatives. The national government's onsite headquarters initiated the meeting, which decided to organize group visits to disaster-affected municipalities with the members who attended the four-party meetings. This decision was made to improve coordination at the municipal level. In April, the group visited Ishinomaki City continually. As a result, the three-party meeting was set up in Ishinomaki City, which stimulated the establishment of a three-party meeting in other cities such as Kesennuma City and Onagawa Town. With respect to hot meal distribution, the meeting set a common objective to provide a hot meal at least once a day to all survivors living in evacuation shelters. They prepared a list to share information about who was working where and, based on this list, they allocated NGOs/NPOs according to their capacity to provide hot meals. In the meantime, the four-party group continued to convene coordination meetings. During these meetings, representatives discussed the topics that were considered to be most pressing; the group then went about setting common objectives to meet the challenges discussed.

Second, local governments coordinated amongst themselves. In Japan, according to the Basic Disaster Act, the loss of municipal capacity is compensated by support from other municipalities. In the Great East Japan Earthquake, municipalities along the Pacific coast suffered severe human and material losses. Prefectural governments tried to gather local resources and coordinate those provided by other prefectural governments, but it was difficult to match the need. This was why other municipal governments that were not stricken by disaster decided to provide assistance to those that were affected. In the Kansai region (west Japan), the Union of Kansai Governments had been established in December 2010, and prefectures that belonged to it decided to provide support by choosing counterparts.

Third, private industry was integrated into the disaster response. After the Hanshin-Awaji Earthquake, private companies became important resources to support survivors. The number of contracts between local government and private companies – especially companies that distributed food, water, daily commodities or medicine – increased. These agreements worked efficiently in the disaster, with each company automatically providing support. Companies voluntarily provided services for survivors using their specializations. Delivery service companies voluntarily distributed supplies from warehouses to evacuation shelters, checking requests from evacuees. As there was no water supply and it was difficult to keep evacuation shelters clean, disinfection supplies were provided by pharmaceutical and hygiene companies. Months after the tsunami, the problem of pest control became serious because many fish-processing industries were damaged and fish

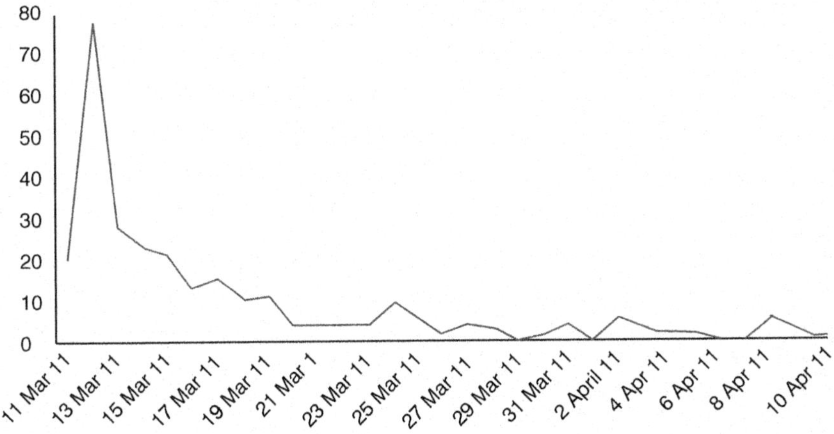

Figure 6.6 Patients transferred to hospitals from evacuation shelters

products were rotting. It took months for the government to set up a new system, but private companies specializing in pest control volunteered their services. Such activities are considered to be corporate social responsibility (CSR), and private industry will be expected to play such a role in future disasters. Thanks to those efforts the living environment in evacuation shelters improved in April. Partition walls were installed by NGOs. Tents were also set up for women to change clothes or nurse babies. There were corners established for children to play.

Despite these efforts, health conditions in evacuation shelters were severe, especially in the few weeks following the disaster. Figure 6.6 shows the number of patients transferred to hospitals from evacuation shelters in Sendai City, Miyagi Prefecture. The number was highest in the first week after the disaster. The experience of the Great East Japan Earthquake revealed that, in the case of a multi-location disaster, it is very difficult to get support immediately after the event. The situation becomes worse when local government suffers. So the preparation efforts of each individual and community are important for survival, especially in the first weeks after the disaster.

Temporary housing

Temporary housing construction was very difficult following the Great East Japan Earthquake because the disaster-stricken areas were along a Rias coastline, a long coastal inlet and narrow plain formed by submergence. Towns developed in the narrow plains were totally wiped out by the tsunami. Afterwards, the government restricted construction of housing in disaster-stricken areas to mitigate future tsunami damage. It therefore became difficult to find proper land for temporary housing. Housing was developed on the few remaining plains such as school grounds, parks or fields. Temporary housing construction started in April 2011 and was mostly completed by the end of October 2011. Based on the experience

Table 6.3 Temporary housing in Miyagi Prefecture

City	*Housing complex*	*Houses (total)*	*Nursing homes for elderly / people with disabilities (PWD)*			*Community centres*	*Residents*
			Elderly	*PWD*	*Total*		
Sendai	19	1,523	18		18	20	1,497
Ishinomaki	131	7,297	88	56	144	114	7,190
Shiogama	7	206				5	194
Kesen numa	93	3,504	45		45	74	3,357
Natori	8	910	16	5	21	7	884
Tagajo	6	373				6	373
Iwanuma	3	384				2	370
Higashi Matsushima	25	1,753	16	10	26	19	1,711
Watari	5	1,126				7	1,094
Yamamoto	11	1,030				9	1,008
Shichigahama	7	421				4	412
Ohsato	1	15					1,272
Misato	2	64				1	2,170
Onagawa	30	1,294	9		9	22	15
Minami Sanriku	58	2,195	27		27	31	62
Total	406	22,095	219	71	290	321	21,609

Source: Compiled from data provided by Miyagi Prefecture.

of the Hanshin-Awaji Earthquake, several attempts to secure human lives were made.

First, several municipalities tried not to separate communities in temporary housing. For example, in Yamamoto Town, Miyagi Prefecture, housing assignments took into consideration both community and school districts. In order to avoid concentrations of vulnerable people in certain housing complexes, the municipality formed groups of households with different categories, selecting 20 per cent from each group. The following are the categories in order of priority.

1 households with elderly (more than 70 years old)
2 households with elderly (more than 60 years old) and households with preschool children
3 households with elderly (more than 65 years old)
4 others

The reason for putting households with people over 60 years and households with preschool children together was that these individuals and families stay in the same area longer. Generally, 60 is the retirement age for Japanese people, and those retired prefer to spend the rest of their lives in quiet and enjoyable circumstances without changing dwellings. Households with preschool children prefer not to change school districts. Thus, mixing both types of households is expected to revitalize the community. Similar efforts not to separate communities and avoid concentrations of vulnerable people were made in other municipalities. However, for several municipalities with limited space, it was difficult to construct temporary housing complexes that could house the entire pre-disaster community. Moreover, it was difficult for municipalities that lost office buildings and staff to go to such elaborate efforts to select residents.

Second, the Ministry of Health sent a notice to prefectural governments on 19 April 2011 that advocated the necessity of community centers to provide social services to residents, and said the government was ready to support such efforts. Table 6.3 lists the temporary housing in Miyagi prefecture. In several complexes, nursing homes for elderly people or people with disabilities were also constructed in the complex. Community centers were installed in almost all housing complexes. LSA service systems, to provide health checks and social networking services for residents, were established in all housing complexes, financed by local government.

Extensive efforts have been made in the recovery phase based on the Hanshin-Awaji Earthquake experience, but several problems remain. The number of disaster-related deaths was worse compared to the Hanshin-Awaji Earthquake: 1,632 as of 31 March 2012. Table 6.4 shows the number of disaster-related deaths by prefecture. The number was greatest in Fukushima, where many people were forced to evacuate because of the nuclear power plant accident. Sudden evacuation meant that patients were obliged to move to different evacuation shelters without proper medical care for hours. These conditions were particularly hard on many elderly people in hospitals. Recovery is ongoing, but no doubt this will be a long road. Even now, 347,000 people live away from where they were before the disaster, still holding onto the hope of returning home.

Table 6.4 Disaster-related deaths by prefecture following the 2011 Great East Japan Earthquake (31 March 2012)

Prefecture	*Dead*	*Missing*	*Disaster-related deaths*	*Disaster-related deaths/deaths (%)*
Iwate	4,671	1,222	179	3
Miyagi	9,517	1,581	636	6
Fukushima	1,605	214	764	42
Ibaraki	24	1	29	116
Chiba	20	2	3	14

Source: Compiled from data provided by the Recovery Agency

Conclusion

This chapter has discussed human security in the disaster response and recovery phases through the experiences of the 1995 Great Hanshin-Awaji Earthquake and the 2011 Great East Japan Earthquake. It revealed that for human security both pre-disaster and post-disaster risk preventive efforts were required; however, less attention was paid to securing people's lives in the disaster recovery process. This caused problematic phenomena such as disaster-related death, isolated death and disaster-related disability during the recovery process following the Hanshin-Awaji Earthquake.

Life recovery was one of the lessons obtained from the Hanshin-Awaji Earthquake. Life is made up of the interaction of several different factors such as housing, social ties, community, health, preparedness, economy and governance. Thus, for life recovery not only housing but also the recovery of other critical factors becomes important. This closely reflects the people-centered approach of human security, which is concerned not simply with the material needs of individuals, but the ways their identities and senses of security are created and maintained through society. Recovery policy should thus integrate people, their societies, safety and lives from the beginning. Recovery policy that separates existing communities and social networks will likely fail.

The Great East Japan Earthquake was a multi-location, multi-hazard disaster and the damage pattern was different from that of the Hanshin-Awaji Earthquake. The Japanese government had developed disaster response plans based on the Hanshin-Awaji Earthquake experience, and thus authorities were not well prepared to deal with an earthquake, a tsunami and a nuclear power plant accident occurring at the same time. The concept of human security is well reflected in the disaster recovery process of the Great East Japan Earthquake. However, the recovery process is still ongoing, and we await further research. There is an especially urgent need to improve current evacuation systems that use school buildings or gymnasiums as evacuation shelters. Attention must also be granted to the issue of shelter management, which should incorporate gender, disability and hygienic support systems.

Japan is an archipelago on top of four different continental plates. The plates have been active recently, and no doubt there will be future mega-disasters. Perhaps an earthquake will directly hit the Tokyo metropolitan area, or a Nankai Trough earthquake-tsunami might occur in the near future. The possibility of a similar disaster somewhere in the world is high. It is important to study in detail the characteristics of each natural hazard and the recovery process to reduce future human casualties, and share experiences with other disaster-prone countries.

References

Kobe City 2000, *Hanshin-Awaji dai-shinsai kobe no seikatsu saiken 5 nen-kan no kiroku* (Five years records of life recovery of Kobe city), Kobe: Kobe City.

Kobe City 2010, *Comprehensive strategy for recovery from the Great Hanshin-Awaji Earthquake.* Kobe: Kobe Institute of Urban Research.

Mizuno, K. 1995, *Hanshin-Awaji junkai rehabilitation team hokokusho* (Report on Hanshin-Awaji Earthquake evacuation shelter rehabilitation team visits). Kobe: Kobe University.

Sakamoto, M. 2011, 'Understanding the living conditions of individuals disabled due to disaster and policy recommendations to prevent future victims', *Journal of Social Security Science*, 15, pp. 395–403.

Sakamoto, M. and Yamori, K. 2009, 'A study of life recovery and social capital regarding disaster victims: A case study of Indian Ocean tsunami and Central Java Earthquake recovery', *Journal of Natural Disaster Science*, 31(2), pp. 13–20.

Shimokawa, T. 2011, 'Koreisha no kodoku-shi no boshi to chiiki mimamori katsudo' (Prevention of isolated death of elderly and support visit activities), *Handbook of Disaster Management*, 3, pp. 174–7.

Tamura, K., Hayashi, H. and Tatsuki, S. 2001, 'A quantitative verification of the seven elements model of socio-economic recovery from the Kobe earthquake', *Journal of Social Safety Science*, 3, pp. 33–40.

Tatsuki, S. 2007, 'Long-term life recovery process among survivors of the 1995 Kobe earthquake: 1999, 2001, 2003 and 2005 life recovery survey results', *Journal of Disaster Research*, 2(6), pp. 484–501.

Tatsuki, S., Hayashi, H., Yamori, K., Noda, T., Tamura, K. and Kimura, R. 2004, 'Model building and testing of long-term life recovery processes of the survivors of the 1995 Kobe earthquake: Structural equation modeling (SEM) of the 2003 Hyogo Prefecture Life Recovery Survey', *Journal of Disaster Research*, 6, pp. 251–60.

Ueda, K. 1996, 'Shinsai kanren shibou to sono taisaku' (Disaster-related death and its prevention), *Nihon Iji Shinpo*, 3776, pp. 40–4.

7 Towards a people-centered housing recovery after the triple disaster

Elizabeth Maly

Introduction

The magnitude 9 Great East Japan Earthquake that struck northeast Japan on 11 March 2011 and the tsunami that followed devastated communities along 500 km of coastline in the Tōhoku region of Japan. Almost 20,000 people lost their lives, and over 470,000 people were evacuated from their homes. As of November 2013, more than two and a half years later, more than 277,000 people are still living as evacuees or in temporary housing (Tasukeai Japan 2013). With the triple disaster of earthquake, tsunami and nuclear accident, the Great East Japan Earthquake (GEJE) is a complex disaster affecting a huge area, including multiple prefectures and municipalities. The impacts of recovery will continue to shape the lives of survivors for years and generations to come. Policies and implementation of housing reconstruction play a pivotal role in the ability of individuals and families to rebuild their lives. The spatial forms and social character of neighborhoods and towns have been altered beyond recognition, and community recovery is closely tied to housing reconstruction as well as town planning. The urban form and local social fabric that will emerge after reconstruction will define both the resilience of individual communities as well as the sustainability of the region overall.

The closely related concepts of human security and people-centered housing recovery offer a way to think holistically about the needs of disaster survivors in the Tōhoku region, and at the same time to consider very specific implications for the form and direction of recovery and reconstruction. Human security explicitly calls for '*people-centred*, comprehensive, context-specific and prevention-oriented responses that strengthen the protection and empowerment of individuals and their communities' (UNSG 2012, author's emphasis). Human security includes a proactive stance towards prevention to reduce the impact and damage from disasters (UNSG 2012, p. 8). In post-disaster recovery, this translates to reconstruction that includes 'building back better', to reduce damage from future disasters. People-centered recovery and human security both support an inclusive recovery – the need to support all people including more vulnerable members of the population – and share a holistic view of the lives of disaster survivors, requiring that economic and livelihood recovery occur in tandem with housing reconstruction.

A people-centered approach can suggest ways to apply a human security approach to specific aspects of housing recovery. After a natural disaster, the loss of stable and secure housing is directly related to the fundamental freedoms of human life with which human security is concerned – freedom from fear, freedom from want and freedom to live in dignity (UNSG 2012). A people-centered housing recovery approach addresses these issues by calling for policies, planning processes and implementation of housing construction that supports residents' abilities to rebuild and recover their lives. This requires ways and options for residents to be able to rebuild on their own, and/or housing support for people who cannot or do not wish to rebuild. It requires flexibility so that survivors are able to choose rehousing options that make the most sense for their own lives, and that are responsive to their individual and household needs, such as income and livelihood. The outcome of a people-centered housing recovery will be residents in stable housing situations, in dwellings that match their physical, cultural, social and economic needs. Driven by policies that support their needs, people-centered housing recovery is accountable to residents.

In the context of post-disaster housing reconstruction in Japan since the 1995 Great Hanshin-Awaji Earthquake, progress towards an increasingly people-centered housing recovery process can be seen. However, the massive scale and complexity of the Great East Japan Earthquake and tsunami, magnified by the regional socio-economic context, means the recovery process is much more difficult than anything Japan has faced in the past. The problems of an aging society and rural depopulation throughout Japan are more extreme in the coastal region of Tōhoku. Beyond ensuring that individuals are able to rebuild their houses and their lives, planning and reconstruction decisions must also consider the long-term sustainability of the region. The triple disaster has also created groups of evacuees who are displaced because of nuclear radiation, and there is a critical and immediate need for policies to support these people who are facing long-term evacuation, life in displacement and complete uncertainty about possible housing reconstruction. Whereas the trajectory of policies developed to support disaster survivors and housing reconstruction since 1995 includes improvements in terms of a people-centered approach, there are major and immediate changes needed to create a people-centered recovery for the survivors of the GEJE.

Based on the human security framework, this chapter considers what is needed for a people-centered approach to housing recovery in Tōhoku, the challenges faced in terms of the vulnerability of disaster survivors and the need to balance disaster risk-based planning with a holistic recovery that includes livelihood along with housing. This chapter also applies a people-centered approach to different aspects of the housing recovery process and specific challenges in the context of Tōhoku, including evacuation, temporary housing, the construction of recovery public housing and individual residents' ability to rebuild. There are also specific housing recovery issues stemming from displacement caused by radiation contamination in Fukushima. A people-centered housing recovery in Tōhoku is not possible unless the Japanese government is held accountable and addresses the housing recovery needs of Fukushima survivors.

A holistic approach to recovery and livelihood

Housing and livelihood recovery are interrelated and proceed together, and the ultimate question for housing recovery is if it supports residents in the struggle to rebuild their lives and livelihoods. The 1994 UNDP report also emphasizes the importance of people having the agency to make their own choices with regard to livelihood in order to have security in their lives (UNDP 1994). Financially, the costs of housing recovery directly impact survivors' livelihoods through their household budgets. Beyond the reconstruction costs that survivors pay directly, relocation and timing of housing recovery have other economic impacts for households, whose individual needs for daily life are intertwined and complex. The location of temporary as well as permanent reconstruction housing is crucial for maintaining activities of daily life (such as school, child care, shopping, hospital access) as well as those directly connected to residents' income generation (workplace or business location). Local business in the disaster area is closely connected to residential factors, and these must be considered together for sustainable town recovery. This is true in general, and also very relevant to the current situation of the towns along the northeast coast of Japan, where local businesses will struggle to survive during and after the multi-year project of residential relocation to higher areas. The effects of the post-disaster recovery process are critical for the livelihoods of survivors, and the human security framework supports a holistic recovery that is sensitive to the protection of livelihoods, along with a reconstruction that reduces the risk from future disasters.

As the 2012 UN Secretary-General's Report stated, human security aims to ensure the 'survival, livelihood and dignity of all individuals', especially the most vulnerable (UNSG 2012). In disaster research, the connection between vulnerability and disaster impact is well-established, as is the way that factors of vulnerability increase the risk and impact of natural hazards (Bolin 2007). The vulnerability analysis is often expressed in an equation as Risk = Hazard × Vulnerability (Wisner et al. 2004). It has been well documented that groups of people who are vulnerable – because of their age, income, gender, disability, racial or ethnic minority status, or housing situation – suffer disproportionately during and after disasters. Without careful support, they will almost certainly be left behind during the recovery process. In terms of housing recovery, low-income residents face additional challenges – especially renters of inexpensive rental housing destroyed by disaster – as do elderly or sick residents. It is critical that recovery policies and programs support all members of the community, especially those who need additional help in order to restore the human security aims of 'survival, livelihood, and dignity' after a disaster.

With regard to housing and the built environment, issues of vulnerability are directly related to socio-economic privilege or lack thereof, and impact residents' living environments, with direct consequences for their housing situations. Simply put, poor people are more likely to live in risky areas and/or in poor-quality housing, and are therefore more vulnerable to natural disasters, both in terms of their own personal safety and health as well as housing/property damage. In addition to economic status, other vulnerabilities – such as being elderly, handicapped, part of a minority group, a single mother, etc. – can multiply challenges for residents. Structural racism and other types of historic and institutionalized discrimination and disenfranchisement also contribute to vulnerability to disaster, as these historic factors shape residents' current situations (Hyndman 2014; Zack 2014).

Human security does not replace the vulnerability framework for disasters, but rather provides a complementary lens through which to consider the same phenomenon. Based on the principle that all residents should have the ability to reconstruct their lives post-disaster, the concept of human security and a people-centered approach both call for the equitable inclusion of all. The direct implication of these principles in practice is that recovery policies must not only support all residents but also target the needs of vulnerable groups and provide appropriate support to ensure their recovery.

Working towards the goal of housing that matches the needs of the residents, a people-centered framework can be applied to housing recovery at three levels: (1) policy at the scale of the disaster area, (2) process at the community scale and (3) housing form and location at the scale of the individual dwelling. At the policy level, decisions about compensation and housing reconstruction have a major and direct impact on whether or not it will meet residents' needs. At the community level, the participation of residents in the decision-making process has a singularly large effect on the degree to which reconstruction decisions reflect the needs of residents. At the individual or household scale, the form and location of the housing used in each phase, starting immediately post-disaster through permanent reconstruction, affects residents' ability to recover their lives. A people-centered approach at all three of these levels is important to create the best possible housing recovery outcome, one in which the resulting housing reconstruction exactly matches residents' needs.

A people-centered recovery is one that matches the needs of residents not only for their immediate dwelling and living environments, but also for their community, social and economic needs. People-centered recovery therefore must encompass the interactions and interrelationships between the design of the physical environment (including both housing and planning) and the social and economic aspects of residents' livelihoods. These relationships are situated within the compressed and contentious timing of the recovery process itself, including the urgency of immediate decisions that affect short- and long-term outcomes, the time needed for planning and carrying out the recovery process, and considerations of the long-term impacts of the recovery process. Although people-centered recovery involves complex needs and considerations, the essence of the desired outcome is clear: people-centered housing recovery will result in houses whose form, design and location match the needs of residents for their daily lives.

Starting from Kobe: the Great Hanshin-Awaji Earthquake and post-disaster housing reconstruction in Japan

Before discussing the challenges and imperatives of people-centered housing reconstruction after the Great East Japan Earthquake, current plans and policies must be considered within the context of the housing reconstruction policy that has evolved since the 1995 Great Hanshin-Awaji Earthquake, the first large-scale urban disaster in Japan since the Second World War. After natural disasters in subsequent years, the adaptations of these policies form a trajectory with some aspects that represent progress towards a more people-centered recovery. Policy adaptations have created improvements in terms of: increased acknowledgment of and support for psychosocial and community needs throughout the recovery process; more efforts to keep communities together and for relocation projects to be closer to former communities; more financial support available for private reconstruction; and more flexibility for residents to choose post-disaster housing options. There have also been improvements in design, construction and livability of both temporary housing and public housing built for disaster survivors. The evolution of housing recovery policies toward becoming more accountable to residents' needs is not enough to guarantee a people-centered housing recovery after the 2011 GEJE, but this trajectory is important for understanding the current situation.

On 11 January 1995, at 5:46am, the magnitude 7.3 Great Hanshin-Awaji Earthquake struck near Kobe City, in Hyogo Prefecture, killing over 6,400 people and destroying over 240,000 houses. After the earthquake, fires broke out and spread quickly through dense low-rise neighborhoods of traditional wooden row houses. The narrow streets and alleyways in these areas were blocked with earthquake debris, and emergency vehicles were unable to enter to extinguish the flames, which burned for days. The areas that suffered the greatest losses of housing to fire were primarily home to lower income and working class residents who relied on the affordable rents of privately owned wooden housing units. In Kobe, as is often the case in natural disasters, this vulnerable population of low-income renters suffered a disproportionate degree of housing damage, and were also at a significant disadvantage through the recovery process, as they were dispersed to distant locations for a long time during the temporary housing phase, and as affordable housing and many small-scale employers in the pre-disaster neighborhood did not return. Public housing was eventually built in this area of Nagata Ward, but much of it took seven years to complete, by which time residents had established their lives elsewhere. In addition, the amount of affordable housing units constructed in this area did not replace the number lost. The pre-disaster housing had been affordable precisely because it was old; once destroyed it is almost impossible to replace with newly built housing with the same level of affordability.

In the areas of Kobe that suffered most from the fires, houses were not the only structures that burned. In these damaged areas of Nagata Ward, one local industry was the construction of 'chemical shoes', a small-scale fabrication process that supported the existence of small and micro-enterprises that assembled parts that were provided to larger companies, who completed the process. Local residents relied on these businesses for employment and livelihood. This local area functioned as an ecosystem for residents, who relied on a system of small-scale, interconnected enterprises for work, and low-cost affordable rental units for housing. When both were destroyed in the fires after the 1995 earthquake, this ecosystem was destroyed. Without additional support, which they did not receive through government recovery projects, very few of these small business owners could recover their enterprises fully, even after years. Similarly, there was limited support for owners of low-cost private rental housing to rebuild.

Temporary housing in Kobe

At the peak after the Great Hanshin-Awaji Earthquake, there were 220,000 displaced residents living in temporary housing (Hirayama 2000). Most of these were single-story steel barracks without soundproofing or insulation, and were uncomfortably hot in summer and cold in winter. The small units (less than 27 square meters each) made for cramped living quarters, and the local environment surrounding the temporary housing areas lacked social facilities and green spaces.

To address the large need for temporary housing, Kobe City used available open land that was slated for development – but this was in inconvenient locations far from services that residents needed for daily life (schools, work, hospitals, shopping) and from their former communities. Residents were awarded temporary housing through a lottery system, which incorporated priority selection for elderly and other vulnerable evacuees such as those with disabilities or families with small children. The lottery system scattered people among different temporary housing locations, disrupting former community ties. Priority selection for vulnerable residents resulted in temporary housing with a disproportionate number of elderly inhabitants, who were cut off from former community-based support systems. Many residents chose not to move into this temporary housing, primarily because of the location that would make it impossible to commute to work or school. They chose instead to stay in emergency evacuation shelters or parks for up to a year after the earthquake, until they were forced to move out (Edgington 2010).

In temporary housing, the serious issue of 'solitary death' emerged, where evacuees passed away without anyone noticing their absence. Although there were some cases of suicide, depression and death related to alcohol abuse, overall the issue of 'solitary death' points to the disruption of any and all social connections or community security. This isolation resulted from the process of assigning residents randomly and a housing form that did not encourage social interaction, and was compounded by lack of social services. The problem of solitary death continued beyond the temporary housing phase, even after disaster survivors moved into high-rise public housing apartments, where again they had no community connections or interactions, and were isolated in high-rise towers located far from original communities in most cases. In the 18 years since the Great Hanshin-Awaji Earthquake, there have been over 1,000 cases of solitary death (Shiozaki 2013). A human security approach recognizes the importance of the connection between community, personal and health security, and the social consequences of housing policy.

Reflecting this connection, in the years since the Great Hanshin-Awaji Earthquake there has been an increased awareness of the need for psychosocial care for disaster survivors. Temporary housing now includes community/social spaces, and social services including regular visits to inhabitants are routine.

Temporary and public housing as part of a one-track housing recovery

Disaster recovery public housing was built in Kobe using government reconstruction funds. Like other public housing in Japan, it is owned by the government, and residents qualify based on income and pay a subsidized monthly rent. These subsidies were initially intended for a limited period of five years, but have been extended repeatedly. Although the concept of the program when established was for short-term support, it is unlikely that survivors in disaster recovery public housing will be able to move out on their own, and the government has no choice but to continue to provide housing for them at very low rents.

Residents faced many of the same issues that they had in temporary housing during the transfer process to disaster recovery public housing, which was located primarily in areas far from former communities and inconvenient for daily life. Again, residents were selected for rental public housing units based on a lottery system, with priority given to elderly or vulnerable residents. The intention to provide extra support to those residents who needed it most is admirable. However, the result again was an overconcentration of vulnerable (mainly elderly) residents. This does not create a sustainable community, and neighborhood relationships of people who relied on one another in their pre-disaster communities were also severed. Especially for the elderly, many of whom had lived in traditional Japanese low-rise housing with a lot of daily interaction on the streets of their neighborhoods, living in high-rise public housing was an extreme lifestyle change.

People who could rebuild on their own did so, using their savings and/or access to private or personal loans. At the time of the Great Hanshin-Awaji Earthquake, the use of government support for disaster survivors for private housing reconstruction was forbidden by law, in accordance with the general principle that the government should not support the replacement of private property. Homeowners without their own resources to rebuild, or renters who were unable to regain stable housing on their own, had no other option for support than to move into disaster recovery public housing. Within this one-track government policy of support for permanent housing recovery, there was no menu of options or different levels of support available to match different housing needs. Eventually, 38,600 households moved into public housing units (Shiozaki 2013). If we consider that some of these people may have been able to rebuild their homes with just a small amount of one-off assistance, it is clear that redirecting some funding to provide multiple options (beyond public housing) could have reduced the need for public housing. In addition, since many residents would probably have been able to rebuild with smaller amounts of financial support than was spent on the construction of public housing, providing this option would be an advantage from the point of view of the efficient use of recovery resources. For both the housing beneficiaries as well as the housing provider, these policies were ineffective.

Support for private reconstruction, owner-driven and self-built housing

There were also positive examples of people-centered housing recovery in Kobe after the earthquake, notably self-built housing, which appeared even during the early temporary housing phase. Although they received no support from the government, Shiozaki found more than 5,000 households had constructed 'self-built temporary housing' in Kobe after the Hanshin-Awaji Earthquake (Shiozaki 2009). 'Self-built' does not necessarily mean that the residents did the construction work themselves, although this was sometimes the case. Residents built their own temporary housing by their own efforts and resources, using prefab or container units, or other structures and combinations. A primary feature of people-centered housing reconstruction is that residents are completely in control of all decisions related to their self-built housing, including cost, schedule, design and location, thus the housing can closely match the residents' self-identified needs. In some cases, these residents were able to continue to live in their self-built housing for the long term, or convert it to permanent housing. Even a very small amount of government support for this kind of reconstruction could have a great impact and, compared to the cost of construction of public housing (at ¥15 million/unit in Kobe), it benefits both the provider and the beneficiary for the government to invest in support for self-built housing (Shiozaki 2013). Moreover, all the government support for the interim housing phase was spent on the construction of prefabricated temporary housing units, which are only used for a few years and then discarded, for which the government pays an additional cost. Some of these prefabricated units were reused in other countries, but there were no options to extend their use by disaster survivors in Kobe. Self-built temporary housing, on the other hand, may be small and/or minimal, yet it offers significant benefits for inhabitants, who are able to stay in their former neighborhoods and rebuild an inhabitable structure quickly. The houses provide smoother transitions for residents, as incremental additions are possible, while continuity is preserved because relocation is not necessary.

Maximizing residents' ability to rebuild on their own allows individual residents to make their own choices, and is an effective way for the housing provider (in this case, the government) to leverage resources to provide more and better housing. As Kondo and Maly demonstrated, involving residents in a more proactive role of 'mobilizing housing' rather than as passive 'housing receivers' leads to more successful recovery results (Kondo and Maly 2013). This parallels recent directions in international development and disaster recovery, which advocate 'owner-driven' housing reconstruction, where residents are in control of decisions about reconstruction of their own houses, and 'cash-based' housing reconstruction, where cash funds for reconstruction are directly provided to the homeowner (Barenstein 2006). Although commonly used in internationally funded housing reconstruction around the world, in Japan there has been strong rejection of the idea of providing government funding to compensate residents for private property, but putting more control of reconstruction in the hands of the residents would give them the flexibility to rebuild and recover in the way that best matches their lives.

Progress toward people-centered policies since the Great Hanshin-Awaji Earthquake

Several aspects of housing recovery after the Great Hanshin-Awaji Earthquake were not people-centered, and recovery policies – such as multiple relocation processes that disrupted community ties, economic and other activities and needs related to livelihood – failed to support residents' life recovery. The one-track housing recovery process in Kobe (prefabricated temporary housing, followed by disaster recovery public housing), and the lack of support for private reconstruction, precluded options for more flexible support matching the needs of individual households. Moreover, weaknesses in design and implementation of temporary and public housing led to long-term displacement and lower quality of life.

In the years since 1995, there have been some policy modifications toward a more people-centered housing recovery in Japan. This trajectory can be seen most clearly in progress regarding Disaster Victim Support between the 1995 earthquake in Kobe and the 2004 Niigata Chuetsu Earthquake. On 23 October 2004 the magnitude 6.8 Chuetsu Earthquake struck a mountainous rural area of Niigata Prefecture, killing 39 people. While the scale of the disaster was much smaller than Kobe, over 100,000 people evacuated from their homes immediately after the earthquake. The number of evacuees steadily decreased over the following weeks and months; 3,460 temporary housing units were constructed, housing 9,649 people at the peak in March 2005 (Iuchi 2010). Local governments tried to avoid repeating 'bitter lessons learned' in Kobe (Iuchi 2010). There were strong efforts to keep communities together throughout the process of evacuation, temporary housing and finally permanent rehousing.

Housing damage in Chuetsu was concentrated in small mountain villages and housing recovery relied primarily on relocation projects, such as Disaster Prevention Group Relocation projects, using national government funding to move groups of residents away from areas of disaster risk. Most residents of mountain villages whose houses had been damaged by the earthquake could choose to have their property included in designated relocation areas, making them eligible for relocation support. As in Kobe, Chuetsu Earthquake survivors had two options for permanent housing recovery: rebuilding on their own or moving into public housing. However, improved government support for both these options made them different from the experience in Kobe, and resulted in a much more people-centered approach to housing reconstruction. Undoubtedly, this was easier to accomplish because of the smaller scale of the disaster and reconstruction projects.

Much of the public housing constructed in Chuetsu was low-rise single family houses built in small clusters. Local governments tried to minimize relocation distance and keep pre-disaster communities together in recovery housing, which often included both private lots for development and public housing for residents who could not rebuild on their own. Designs for public housing were similar to market rate housing, including some innovative examples of wooden public housing. Residents who had moved collectively to evacuation centers, and then into temporary housing, moved into the public housing together with their pre-disaster neighbors. Compared to the number of residents who had stayed in temporary housing, only a fraction eventually moved into public housing. This is primarily because of the additional support for private reconstruction, one of the major factors of people-centered housing recovery in Chuetsu.

One of the most significant differences for homeowners after the 2004 earthquake in Chuetsu compared to the 1995 earthquake in Kobe was the legal change that allowed financial support for disaster survivors from the government to be used in ways related to housing reconstruction. Modifications to the Disaster Victims' Support Act and the 1999 creation of the new Victims Life Recovery Support Act meant that, after the Chuetsu Earthquake, households with completely destroyed houses could receive up to ¥3 million ($30,000), which they could use for reconstruction. In addition to this funding from the national government, Niigata Prefecture also created a complementary program, which was not restricted by age or income and which provided support that could be used for rebuilding or rent (Iuchi 2010). Other examples of local governments supplementing national government support can be seen in the case of the 2000 earthquake in Tottori and the 2007 Noto Peninsula Earthquake (Maki 2007).

Vulnerability in Tōhoku

The Great East Japan Earthquake in 2011 occurred 16 years after the Great Hanshin-Awaji Earthquake. The lessons from Kobe, including the shortcomings of the housing recovery process and the long-term outcome of recovery, were well-known and understood by the various actors involved in the current recovery planning process. Plans for housing recovery in Tōhoku include some aspects that show an increased awareness of housing issues related to the human security of the residents, and policy innovations that show a move in the direction of a more people-centered housing recovery. However, the complexity and massive scale of this disaster, along with other pre-existing socio-economic conditions in the disaster-affected region, means that a long and difficult recovery process is unavoidable, and there are also many issues for concern.

Long before 11 March 2011, the Tōhoku region was already suffering from population decline, with an aging population and outmigration of the younger generation to larger cities. The tsunami and evacuation sped up this process of depopulation. Demographic changes such as these impacted the way the disaster unfolded, and continue to shape future recovery. When the disaster occurred, many of the victims were elderly people who were not able to evacuate safely. Some elderly people also passed away during the evacuation process, especially in cases where people in nursing care facilities were evacuated to escape radiation after the nuclear accident at the Dai-ichi power plant in Fukushima. The disruption and uncomfortable living situation in evacuation shelters, and subsequent relocation to temporary housing, was and is harder on the elderly. As was the case in Kobe, elderly evacuees are seriously impacted by the loss of community connections, which makes it difficult to meet their needs to continue their daily lives.

Thinking ahead to town recovery and housing reconstruction, the regional context of an elderly, aging and shrinking population in Tōhoku has a number of serious implications for recovery and reconstruction. Beyond depopulation, the fact that many of the disaster survivors are elderly has direct consequences for rebuilding. As was the case after the Kobe earthquake, many residents of disaster recovery public housing will likely be elderly, requiring additional support. At the same time, it is important to consider the long-term future of the region and the implications of investing in the construction of a large number of public housing units and the related necessary infrastructure. Even in the best-case scenario, future predictions show a slowed population decline. Recovery plans tend not to address the future population decline in this region – focusing rather on reconstruction for the current inhabitants. The immediate needs of each disaster survivor must be addressed, but this must occur within the context of and in balance with consideration for the long-term sustainability of the region.

Evacuation and displacement

The people-centered framework of human security is fundamental to the evacuation phase after a disaster. Human security calls for a living environment that provides dignity, safety and security, even during the interim phases of evacuation. For the immediate phase post-disaster, evacuation shelters must be located in safe areas, with adequate emergency supplies and living conditions for evacuees. After the GEJE, there were issues and challenges related to the safety of evacuation places and standards of conditions in shelter life, but within the context of housing recovery, the pivotal issues are related to displacement in evacuation.

Within an international framework, evacuees who have been scattered throughout Japan can be considered to be internally displaced persons (IDPs). The rights of and problems faced by IDPs are often minimized or ignored in wealthy 'developed' countries, such as after Hurricane Katrina in 2005 in the United States as well as after the GEJE. Evacuees, especially those from Fukushima, have faced and continue to face a number of issues, such as discrimination and rejection from host communities, and difficulties in receiving benefits from TEPCO, including the challenges of dealing with the complicated paperwork required. People who were evacuated from mandatory evacuation zones because of radiation, and people who chose to evacuate even though their homes are outside the mandatory evacuation area, are not entitled to the same benefits and compensation paid by TEPCO.

Japanese laws that provide compensation for disaster victims specify those affected by natural disaster; therefore residents who left their homes because of radiation cannot receive the same housing benefits for disaster victims and support for recovery as those whose housing was destroyed by the earthquake or tsunami. To support an equitable recovery for all survivors, this policy should be changed immediately, so that nuclear evacuees have the same support to reconstruct their lives. In the cases of residents affected by the nuclear accident, their evacuation and planned return timeline varies by location, leaving many stuck in limbo, waiting to see if or when they can return to their former towns.

Temporary housing

A wide range of temporary housing was used after the GEJE, with large variations in terms of quality, type of construction, scale of housing site, distance from original residential area and degree to which residents were collectively relocated. There are many examples of temporary housing that are a vast improvement over that of Kobe, but also many examples that are no better and which repeat the same mistakes. The large number of evacuees and limited time and land for construction contributed in some part to limited improvements.

Two of the key innovations for the temporary housing phase after the GEJE have been the use of private rental apartments as temporary housing, and the construction of high-quality wooden temporary housing. With regard to private rental apartments, this is the first time this system of 'designated temporary housing' or *minashi kasetsu juutaku* has been used in a Japanese post-disaster context. In this system, the national government pays up to ¥60,000 per month (US$600) for rent for the evacuee household. The system requires a contract between the government, private landlord and resident (evacuee). This system of 'designated temporary housing' has a number of benefits. In areas with housing vacancies, it is quick and easy for evacuees to move into available apartments. It is cheaper for the government to pay rent instead of constructing new temporary housing. However, this system can only be used in areas with available housing stock that survived the tsunami. Therefore, the 'designated temporary housing' strategy is used more near the large city of Sendai and in the surrounding area.

The system is complicated because of the three parties involved in the rental agreement. Because it is a new system, there is some uncertainty and there have been modifications as the program has developed. For now, during the period while evacuees are still living in temporary housing (as of Fall 2013, this limit has now been extended to four years), the 'designated temporary housing' system is also under the same timeline. One of the unresolved questions is what will happen at the end of the temporary housing phase. Because the residents become scattered it is difficult to track these evacuees and provide information and services to them. The use of designated temporary housing may also contribute to an increased or more rapid population transfer to urban areas, as evacuees move away from rural and coastal disaster areas to areas where there is housing stock available for designated temporary housing. These problems notwithstanding, providing the option of designated temporary housing offers a more comfortable living environment for the residents, who are living in housing that has the same level of quality and comfort as permanent housing. In addition, the ability of residents to choose their own apartments means that designated temporary housing supports residents' abilities to choose housing that matches their needs – a key aspect of people-centered housing.

Even with the introduction of designated temporary housing on a large scale, after the GEJE there was an immediate need for a large number of emergency temporary housing units. More than 50,000 units of temporary housing were newly constructed, by a variety of companies. Prefabrication building associations had pre-existing contracts in place with the disaster-affected prefectures (and all prefectures in Japan) before the GEJE, to allow for the rapid construction of temporary housing after a disaster. The quality of temporary housing varied widely, from some very poor-quality housing built by prefab construction companies who were not used to building housing, to higher quality temporary housing units constructed by *housemakers* (Japanese companies who specialize in suburban tract house development), or using timber construction.

The use of wooden temporary housing in Tōhoku is an example of a significant improvement in housing quality for residents, as well as the use of local labor and materials. Wooden temporary housing was used in small quantities in some communities in Iwate Prefecture, such as Rikuzentakata and Sumita Town, whose mayor had a plan in place before the earthquake to use local timber in case of disaster. Fukushima Prefecture made a commitment to use wooden temporary housing on a large scale (more than 4,000 units), partly due to the understanding that residents may be forced to live in these units longer than usually expected for temporary housing. There is substantial variety with different types of wooden temporary housing, such as size and design, but in general it provides a more pleasant living environment. Most wooden temporary housing units are detached, and whereas the long-term prospects for these units is not known, they at least have the potential to be reused in another configuration or for another purpose, as opposed to prefabricated temporary housing, which will likely be discarded after its lifetime of several years.

Public housing and relocation

As in the recovery after the earthquakes in Kobe and in Chuetsu, public housing will play a significant role in housing recovery after the GEJE. As of October 2013, plans for new public housing in the three most devastated prefectures call for the construction of over 6,000 units in Iwate Prefecture, 15,000 units in Miyagi Prefecture, and between 3,000 and 4,000 units of public housing in Fukushima Prefecture (Japan Reconstruction Agency 2013). The need for public housing as a form of social support for residents is unavoidable; and for the many disaster survivors who are unable to rebuild on their own, the provision of public housing is absolutely necessary for them to regain stability in their daily lives. However, the construction of public housing in Tōhoku is complicated because of several aspects: (1) the location of public housing as part of relocation planning, (2) the calculations for the number of units needed based on the expressed desires of future residents and (3) the long-term sustainability and management of the housing.

Local governments surveyed disaster survivors to calculate the number of public housing units needed, but compared to these earlier surveys, later surveys show a sharp decrease in the number of people who want to move into public housing, making it necessary to revise the number of units constructed. Initial interest in public housing was likely larger than the actual need, as residents felt it was a good idea to sign up for public housing just in case, although they hoped for an alternative housing solution. On the other hand, as time passes, many residents who had hoped to rebuild on their own may give up on this option, and choose instead to move into public housing. As a fallback option, the number of residents who move into disaster recovery public housing will depend on what kind and how much support is available for private reconstruction.

As most of the recovery planning for towns in coastal areas relies on relocation of residential areas to high land, plans for the construction of public housing have been created in parallel with the designation of new residential areas. Based on experience from previous disasters in Japan and the demographics of the disaster area, many people who choose to move into public housing will be elderly residents who do not have the desire or ability to rebuild on their own. In the case of the recovery after the Chuetsu Earthquake, most relocation moved residents from areas deep in the mountains to more convenient locations near towns and services, making post-relocation lifestyles more convenient in terms of access as well as reducing the burden of dealing with heavy snowfalls. In Tōhoku, on the other hand, relocation will move residential areas away from the sea and the town centers, to new residential areas at higher elevations, making them less convenient and attractive to residents, especially the younger generations who are already moving away to larger towns. Without opportunities for local employment, these relocation areas will be even less desirable, and many people who can choose another option will not participate in the relocation. This could result in high-land relocation areas dominated by many elderly residents in public housing. Not only would this kind of area lack the vibrancy of a multi-generational and mixed-use community, but after 20 years, most elderly residents will have passed away, leaving behind only empty public housing. Beyond the construction of public housing itself, relocation projects require a huge expenditure for the creation of new infrastructure. This massive investment of money and resources should not be squandered on creating relocation areas with no hope for long-term sustainability.

One of the pivotal issues for recovery after the Great East Japan Earthquake is the need to balance safety and relocation with the livelihood needs of residents and communities. Most municipal recovery plans in the disaster area are based on the guidelines provided by the national government, which call for the construction of massive levees, coupled with relocation to higher land areas. This strategy is based on maximum risk avoidance, as residential zones are moved completely away from tsunami-inundated areas, and the coast is protected by massive seawalls. Human security has a clear relationship to risk and preventing damage from natural disasters, and at first glance these recovery policies seem to be doing exactly that. However, considered in a more holistic way, those measures that focus only on risk prevention also have negative effects on local communities. Massive levees will completely destroy the historical landscape and attractive seascape of the towns. Also, local people have lived together with the sea for generations, and such developments will affect this relationship. As Bird describes, the massive levees will also damage the coastal ecosystem (Bird 2013). The relocation of large parts of coastal residential areas will also have a significant impact on these towns' futures. During the years required for relocation and reconstruction in the new high elevation areas, there is a serious question whether and to what extent residents will be able to have continuity in their daily lives. If disaster recovery is to be genuinely people-centered, it is important to adopt a holistic point of view on the impact of recovery planning and reconstruction on the lives of disaster survivors. Other issues besides risk mitigation are important to disaster recovery, and it is also important to consider the importance of long-term community security and livelihood sustainability.

For the residents of Fukushima who have evacuated from radiation, the issue of relocation and community sustainability is even more complicated. Within the area contaminated by radiation, there are some areas where residents may return to live; areas where the government is planning to allow residents to return after decontamination; and areas where the high levels of radiation mean there are no current plans to allow residents to return. People whose homes are in towns in any of these categories face uncertainty about the future, and are waiting for the next phase of housing recovery, which in any case will be delayed compared to the recovery planning and housing construction in neighboring Iwate and Miyagi Prefectures.

Unlike the areas that suffered tsunami damage, towns that suffered the highest amounts of radioactive contamination have relocated completely, including the town administrative offices, and are functioning within host towns. In these cases, there are also plans to build public housing for the evacuees in the host towns. In other cases where radiation levels are lower, the government is planning to construct public housing within those towns. However, in these areas, the danger of radiation is preventing younger families from returning, as families with small children will do whatever they can to avoid the risk of living in an area with radioactive contamination, even if the level is judged 'safe' by the government. In these cases, the policy to allow residents to return, and to construct public housing contributes to the fracturing of communities, as only the elderly will return. In the areas whose fate has not yet been decided, residents are stuck in limbo, which also leads to a split as residents who can rebuild on their own, or younger working age families or those with school-age children, make their own choices about where to move to create stability for their own households. If the government were to decisively announce that residents of certain areas will not be allowed to return to former towns, and instead to create a designated area for them to relocate to, it might be possible to keep parts of former communities together post-disaster. However, as official policy is just that 'return to former towns is just not possible *yet*', there are no such plans to relocate residents to new areas. By taking no action, these communities will continue to fragment and residents will scatter, leaving only the most vulnerable (elderly) who will remain in temporary housing for an extended time, waiting to see what will happen to them in recovery.

Homeowners in Fukushima whose houses were directly damaged by the tsunami can receive the standard amount of Disaster Victim Assistance based on the level of housing damage (up to ¥3 million for a completely damaged house). However, people who cannot return to their houses because of radioactive contamination are not currently eligible for the same assistance based on housing damage. At the same time, even abandoned houses that were not damaged by the tsunami will continue to fall into disrepair when exposed to the elements. Recently, government plans have been announced to provide for evacuees from Fukushima to purchase new housing in the areas that they have evacuated to, as the government has finally abandoned their policy that everyone will eventually be able to return to their former towns. This policy should be immediately expanded so that these evacuees from Fukushima can receive the same support for housing reconstruction as other GEJE survivors. There is a clear precedent here, the 'designated completely destroyed' category used after the 2007 Chuetsu Earthquake, for houses which were damaged by the heavy snowfall during the winter after the October earthquake.

Avoiding a one-track path and supporting options for housing recovery

As discussed above, after the 1995 Great Hanshin-Awaji Earthquake, the housing recovery policy followed a 'one-track' path: prefabricated temporary housing followed by public housing. The government took responsibility for housing disaster victims and adopted a principle of social welfare support, but there was only one inflexible option provided, with no compensation available for private homeowners to rebuild.

In the years since the Great Hanshin-Awaji Earthquake, disaster recovery policy changes in Japan have led to more flexibility and support for residents' private reconstruction. This support from the national government – combined with additional sources of support – may be enough to support residents' reconstruction. As seen in the previous examples after other recent disasters, if this combined amount is large enough, residents will be able to leverage it for rebuilding on their own. The basic funding available to owners of completely damaged houses as Disaster Victim Support, ¥3 million per household, is not available to owners of partially damaged houses or for nuclear evacuees, who may therefore have a much harder time repairing their houses if they do not have the personal resources to do it on their own. From a human security perspective, the most fundamental and critical aspect is whether the *options for housing reconstruction* match the needs of the residents, and whether they have adequate choices and the opportunity to select support which is suitable to their own housing needs and desires.

Conclusion

Looking back at how post-disaster housing recovery policies in Japan have evolved since the 1995 Great Hanshin-Awaji Earthquake, we can see a progression towards a more people-centered approach. These improvements in housing recovery policy and implementation especially benefited survivors after the 2004 Chuetsu Earthquake. Significant attempts to increase the human security of disaster survivors during the housing recovery process have focused on avoiding the mistakes of housing recovery in Kobe: poor-quality temporary housing; scattered communities; remote locations for temporary and public housing; housing forms and designs that lead to social isolation; lack of support for elderly/vulnerable residents; a single-track housing recovery process; and a lack of support for private reconstruction.

Housing recovery policies after the Great East Japan Earthquake include many aspects that are great improvements over the experience of Kobe. There are examples of excellent initiatives that support human security and people-centered approaches to housing recovery, such as: the provision of social support staff; community spaces and activities in temporary housing; attempts to keep communities together; high-quality/wooden temporary housing; and planned small-scale public housing. Yet there are also many examples of the same mistakes from Kobe being repeated in Tōhoku, such as poor-quality temporary housing and scattered communities. In addition there are new and massive challenges for housing recovery in Tōhoku, such as residents facing displacement and relocation because their homes are contaminated by radiation, and a disaster area that was already facing regional economic decline and depopulation. Therefore, although Japanese policies have improved, there are significant and urgent changes that must be made for the housing recovery after the Great East Japan Earthquake to become more fully people-centered.

In terms of housing reconstruction as well as long-term recovery issues, the fate of the evacuees who fled radiation in Fukushima is still unknown, as many are stuck in a long-term waiting pattern, uncertain as to when they will be able to return. Many are choosing to evacuate or not to return, especially families with small children. The loss of this diaspora will have a substantial long-term impact on the emptying towns left behind, and recovery policies must shift towards a more holistic and people-centered approach, to consider options for creating sustainable future communities in this area. There is a risk that entire towns and public housing will be rebuilt solely for elderly residents, and that many residents in temporary housing face long-term uncertainty, not knowing whether or when they will be able to return to their home towns. A better approach might be to proactively create new relocation areas, to allow residents to move on and start to rebuild their lives. Besides the problems associated with location, and with the construction of public housing, current policies related to private housing reconstruction do not adequately support residents whose homes have been damaged, or are uninhabitable because of radiation. For the human security of these homeowners, they should be entitled to the same support for rebuilding, whether their house was directly damaged by the tsunami or made uninhabitable by radiation.

A people-centered approach to housing recovery is a way to apply these principles specifically to the realm of housing recovery and reconstruction, which is directly connected to human security's concern with 'how people live and breathe in a society, how freely they exercise their many choices, [and] how much access they have to market and social opportunities' (UNDP 1994). Human security and a people-centered approach to housing recovery are overlapping concepts that can support and reinforce each other. Human security provides a framework that can bridge issues related to disaster risk, prevention and recovery planning, along with life recovery, to consider how to be better accountable to the needs of disaster survivors throughout the recovery process. The 1994 UNDP report explained that the loss of human security 'can be human-made – due to wrong policy choices. It can stem from the forces of nature. Or it can be a combination of both – as is often the case when environmental degradation leads to a natural disaster, followed by human tragedy' (UNDP 1994). The loss of human security experienced by survivors of the GEJE was and is certainly a result of this combination of man-made policy and natural disaster – although it can be argued that there is no such thing as a purely 'natural' disaster, since damage is always the result of the interaction between a natural hazard event and the physical, built and social environments created by people (Perry 2007, p. 9). However, the quotation from the UNDP report correctly emphasizes the importance of addressing policy, the forces of nature and their interaction, toward the goal of preventing the loss of human security of disaster survivors.

Having a safe dwelling is a fundamental precondition for human beings to feel secure, and addressing housing security is one critical part of post-disaster life recovery. A people-centered approach to housing recovery offers a way to apply the concept of human security to housing reconstruction at multiple levels: from the overall housing recovery policy, the planning process at the community level, down to the design and location of individual dwelling units. People-centered housing recovery offers a way to think about housing recovery at these different levels so that together they create housing that is appropriate for and based on the needs and desires of the residents, to best support their life recovery.

References

Barenstein, J.D. 2006, *Housing Reconstruction in Post-Earthquake Gujarat*, Humanitarian Practice Network Paper No. 54, London: Overseas Development Institute, http://www.odihpn.org/documents/networkpaper054.pdf, viewed 16 March 2014.

Bird, W. 2013, 'Post-tsunami Japan's push to rebuild coast in concrete', *Asia-Pacific Journal*, 11/21(1), 26 May, http://www.japanfocus.org/-Winifred-Bird/3945#sthash.dAGQUYaz.dpuf, viewed Nov. 2013.

Bolin, B. 2007, 'Race, class ethnicity and disaster vulnerability', in H. Rodriguez, E. Quarantelli and R. Dynes (eds), *Handbook of Disaster Research*, New York: Springer, pp. 113–129.

Edgington, D.W. 2010, *Reconstructing Kobe: The Geography of Crisis and Opportunity*, Vancouver: University of British Columbia Press.

Hirayama, Y. 2000, 'Collapse and reconstruction: Housing recovery policy in Kobe after the Hanshin Great Earthquake', *Housing Studies*, 15(1), pp. 111–28.

Hobson, C., Bacon, P. and Cameron, R. (eds) 2014, *Human Security and Natural Disasters*, London: Routledge.

Hyndman, J. 2014, 'Human security in the face of dual disasters', in C. Hobson, P. Bacon and R. Cameron (eds), *Human Security and Natural Disasters*, London: Routledge, pp. 111–26.

Iuchi, K. 2010, *Redefining a Place to Live: Decisions, Planning Processes, and Outcomes of Resettlement after Disasters*, Urbana, IL: University of Illinois at Urbana-Champaign.

Kondo, T. and Maly, E. 2013, 'Housing recovery by type of resident involvement: Providing housing vs. mobilizing residents', in *Selected Papers from ISHED Conference 2012*, Shanghai: International Society of Habitat Engineering and Design, pp. 73–80.

Maki, N. 2007, 'How can the public sector support recovery of privately owned individual housing after natural disaster? Possibility of setting housing recovery grant in Japan', 2nd International Conference on Urban Disaster Reduction, 27–29 Nov.

Murosaki, Y. 2007, 'Lessons on reconstruction strategies from the Great Hanshin-Awaji Earthquake', *Journal of Disaster Research*, 2(5), pp. 330–4.

Olshansky, R. B., Johnson, L. A. and Topping, K. (with Murosaki, Y., Ohnishi, K., Koura, H. and Kobayashi, I.) 2005, 'Opportunity in chaos: Rebuilding after the 1994 Northridge and 1995 Kobe Earthquakes', Department of Urban and Regional Planning, University of Illinois (web-published March 2011), http://www.urban.uiuc.edu/faculty/olshansky/chaos/Opportunity-in-Chaos-March2011.pdf, viewed 16 March 2014.

Perry, R. 2007, 'What is a disaster?' in H. Rodriguez, E. Quarantelli and R. Dynes (eds), *Handbook of Disaster Research*, New York: Springer, pp. 1–15.

Japan Reconstruction Agency 2013, https://www.reconstruction.go.jp/topics/main-cat1/sub-cat1-12/20131030_sumainokouteihyou.pdf, viewed Dec. 2013.

Shiozaki, Y. 2009, *Jūtaku fukkō to komyuniti* (Housing Recovery and Community), Tokyo: Nihon Keizai Hyoronsha.

Shiozaki, Y. 2013, 'Reconstruction of permanent housing in Tohoku', presentation at technical workshop 'Learning from Mega-disasters', Kobe, 24 June.

Tasukeai Japan 2013, http://tasukeaijapan.jp, viewed Dec. 2013.

UNDP 2011, Annual Report 2010/2011, *People-Centered Development: Empowered Lives. Resilient Nations*, New York: UNDP.

UN-Habitat 2010, *Shelter and Housing: UN-HABITAT in Disaster and Conflict Contexts*, Geneva: UN-Habitat.

World Bank 2010, *Safer Homes, Stronger Communities: A Handbook for Reconstructing After Natural Disasters*, Washington, DC: World Bank.

UNSG 2012, *Follow-up to General Assembly resolution 64/291 on human security: report of the Secretary-General*, UNSG A/66/763, 5 April, https://docs.unocha.org/sites/dms/HSU/Publications%20and%20Products/Reports%20of%20the%20Secretary%20General/A-66-763%20English.pdf, viewed July 2013.

Wisner, B., Blaikie, P., Cannon, T. and Davis, I. 2004, *At Risk: Natural Hazards, People's Vulnerability, and Disasters*, New York: Routledge.

Zack, N. 2014 'The ethics of disaster and Hurricane Katrina: Human security, homeland security, and women's groups', in C. Hobson, P. Bacon and R. Cameron (eds), *Human Security and Natural Disasters*, London: Routledge, pp. 57–73.

8 An ageing society and post-disaster community security

Junko Otani

The UNDP's 1994 Human Development Report outlined seven core components of human security. In the report, natural disasters were classified as threats to 'environmental security'. However, the other six dimensions of human security (health, food, economic, personal, community and political security) have relevance as a framework for understanding the challenges which confront the survivors of natural disasters. Human security is important to the study of natural disasters as it highlights the onset of a range of factors that negatively affect those who are vulnerable or exposed to such events.

Natural disasters can also combine with human-made disasters, which often worsen the degree of damage. When survivors suddenly lose their houses and workplaces, issues related to environmental, food, health and economic security immediately arise. After the initial period of emergency medical relief, health issues continue for the medium and long term. The seven components of human security are highly interconnected. This chapter demonstrates the interconnected nature of human security threats by examining the impact of natural disasters on the health of older people. It has been shown that declining trends in health security were found among those who experienced personal losses and high community destruction (Phifer et al. 1988). Community security plays a key role in supporting health security in the long term, especially for older people. The concept of 'social capital' has been applied in discussions on the vulnerability of victims, as well as empowerment for post-disaster life reconstruction (Aldrich 2012), and is closely related to community security.

Gender is also an important dimension of human security, which needs to be considered when looking at post-disaster community security (Enarson 2014) and the elderly. To demonstrate these themes this chapter focuses mainly on community security, and looks principally at the 1995 Great Hanshin-Awaji Earthquake, which occurred in an ageing society. The cases of the 2011 Great East Japan Earthquake and tsunami, the 2011 Christchurch Earthquake in New Zealand, the 1999 Chi-Chi Earthquake in Taiwan and the 2008 Sichuan Earthquake in China will be used as additional support for the central argument of the chapter.

When examining disasters around the world, older people rarely make the headlines. Japan's media, however, often feature older people in disaster coverage. This was the case with the post-disaster media coverage of the 1995 Great Hanshin-

Awaji Earthquake (Otani 2010, p. 1). Many media commentators reporting on the 2011 Great East Japan Earthquake and tsunami noted that a distinguishing feature of this particular disaster was the number of older people affected (Cunningham 2011). It was not long, however, before the media's attention shifted to the Fukushima nuclear power plant incidents and the proportion of coverage featuring the earthquake and tsunami-affected areas decreased, and consequently so too did the news featuring older survivors. In the case of these disasters it was not so much the immediate effects that disproportionately affected older people. The skewed death toll was due more to a number of secondary consequences. In the 1995 Great Hanshin-Awaji Earthquake, the majority of older people died because they were sleeping in ground-floor rooms and were injured due to toppling furniture and walls. The situation in the 2011 Great East Japan Earthquake and tsunami was different, where many older people survived the earthquake but were unable to escape the tsunami and subsequently drowned.

Ageing society

One of the major features of the 2011 Great East Japan Earthquake and tsunami was that it affected areas with a large number of elderly (Yamamoto et al. 2011). 26.8 per cent of Iwate Prefecture's population was 65 years old or older in 2009 (and is projected to be 37.5 per cent in 2035). At the evacuation centre where the Nagasaki University medical relief team assisted, out of 221 evacuees at the shelter on 28 March, 84 (38 per cent) were people aged 65 years and over. These figures are high, even for a nation with one of the world's oldest populations. In the 1995 Great Hanshin-Awaji Earthquake, more than half of those who died were aged over 60 (Tanida 1996), while 60 per cent of those killed were women. All of the most common injuries and earthquake-triggered illnesses which were reported disproportionately affected people aged over 65. This compares with the overall proportion of those aged over 65 for Japan, which in 1995 was 19.6 per cent, 22.7 per cent in 2000 and is projected to be 36.3 per cent in 2025 (Otani 2010, p. 3). People aged over 60 made up 17.8 per cent of the population in the affected area at the 1995 Great Hanshin-Awaji Earthquake (Otani 2010, p. 14). However, the proportion of people aged over 65 in temporary shelters in Hyogo Prefecture was 30.3 per cent in February 1996, one year after the earthquake, which is roughly twice as many as might be expected (Otani 2010, p. 84). It is necessary to break this figure down further: for example, at a temporary shelter which consisted of three compounds, the proportion of those aged over 65 was 44.2 per cent, 26.3 per cent and 7.1 per cent respectively (Otani 2010, p. 84). The difference in these figures reflected the process of temporary shelter allocation: the higher proportion of elderly people in the first compound might reflect the fact that priority allocation was given to elderly people (Otani 2010, pp. 90–1).

China is another country with an ageing population. The 2008 Wenchuan Earthquake in Sichuan, China, exposed the difficulties of older survivors left alone in mountain villages damaged by the earthquake (Chan 2008; Otani 2010, p. 186). As a result of current economic development policy in China, younger people in rural areas have moved to cities for job opportunities and older people in rural areas tend to be left alone. The earthquake resulted in further migration to urban areas by younger people. It is clear that young people, in this context, were motivated by loss and the search for financial opportunities to cover rebuilding costs. Hence the earthquake led to greater economic insecurity among affected populations. When rural populations migrate in search of employment opportunities and cash income, the fabric of rural communities breaks down. This leads to a weakening of the social capital supply cycle of the rural community in question. This economic insecurity, which leads to rural–urban migration, results in higher levels of vulnerability for remaining populations, and a lessening of community resilience to future natural disasters. Economic insecurity and out-migration therefore cause population insecurity as well as community insecurity. Older people face difficulties when a community ceases to generate a caring environment that enables them to pursue a decent home life.

New Zealand also has an ageing population, despite an influx of migrants. A series of earthquakes from September 2010 had cumulatively caused damage and undermined community capacity to respond when a major earthquake struck central Christchurch in February 2011. Since the quakes, younger, more geographically mobile people are migrating to other parts of New Zealand and to Australia. This leaves older people and other vulnerable groups at risk of being left alone or lacking vital support mechanisms.

Following the disaster it seems that the specific needs of the elderly have largely been overlooked. Sally Keeling, President of the Canterbury Branch of the New Zealand Association of Gerontology, notes that *The Christchurch Press* was one of the exceptions, writing a series of articles on older people, with the headlines alone making for grim reading; 'Some old folk isolated', 'Choice gone for Christchurch old folk' and such. She also refers to 'rest home refugees', stressing the vulnerability, limited autonomy and loss of security facing older people (Keeling 2011). Over 300 rest home residents were part of an emergency evacuation outside Christchurch; in total 600 beds have been lost in the Canterbury aged care sector.

Keeling's piece also introduces more positive stories of community empowerment and individual resilience. The article 'Warmth of a hug and food ease fears' describes a church hall being used as a drop-in centre for older residents in the neighbourhood to come together for shared food and social interaction. Many of their neighbours have moved out, either short or long term, leaving older people more likely to be 'home alone' in their street. Weakening family ties is one cause of elderly vulnerability (Otani 2010, p. 31). Although the destruction of housing and infrastructure was the most visible result of the disaster, there is more to housing loss than the architectural fabric. The destruction of homes also affects community ties and networks (Otani 2010, p. 33). This again highlights the special attention that needs to be paid and programmes that need to be carried out in the response to disaster. Community response networks need to be reinforced in the face of massive public health issues with water, power, sewerage, winter cold and summer heat. All communities in earthquake-prone areas need to consider safety preparedness and mutual support provision after an earthquake has occurred. Clearly further investigation of the problems of ageing societies and post-disaster community security is needed.

Vulnerability and exposure faced by older people

Natural disasters bring to the surface the poorest and most vulnerable groups in society: orphans, the elderly and the disabled (Otani 2010, p. 18). They make visible those who were previously invisible in the everyday world (Varley 1994; Otani 2010, pp. 18, 31). Subsequent studies of mortality patterns from Hurricane Andrew, which hit southern Florida in the USA in 1992, revealed that during the storm older adults were at disproportionate risk of dying from 'mechanical asphyxia' in their homes and from falls. Older people were more likely than younger to die from cardiovascular causes after the storm (Barusch 2011). Morrow (1999) sees the vulnerability of victims to disasters as socially constructed. She identified, taking her study of Hurricane Andrew as a case, the poor, the elderly, female-headed households and new migrants as the categories of people at greatest risk during a disaster response (Otani 2010, p. 31).

The 1995 Great Hanshin-Awaji Earthquake impacted family structures and living arrangements, with many health problems worsening for older people. Some families who thought that the evacuation centres were too uncomfortable for their older members sent them to stay with relatives outside the prefecture or to aged care institutions (Otani 2010, pp. 86–8). The available temporary shelters and permanent reconstruction housing in later years was either too small to enable a large family to live together, or too problematic due to the creation of stressful living conditions (Otani 2010, p. 127). In addition, some families could not afford to house their impoverished parents (Otani 2010, p. 89), instead leaving them in evacuation centres, where at least they would receive aid and support from public health nurses. The middle generations often suffered from the burden of double loans, having to pay off mortgages on their destroyed homes at the same time as they incurred rebuilding costs. This situation was made even more difficult as a result of post-tsunami job loss. The older generations do not want to be a burden for their sons and daughters in a difficult post-disaster context. It is surmised that this destruction of family living arrangements and the loss of cross-generational cohabitation could also lead to a deterioration in the health of family members. This is reflective of the way in which different kinds of vulnerability interact.

With regard to the 3/11 disaster, the elderly were again one of the most vulnerable groups. A high proportion of causalities were the elderly: 92 per cent of deaths were due to drowning in the tsunami, and 65 per cent of those who died were aged over 60. In addition, older people confronted difficult conditions in both evacuation centres and their original accommodation. The disaster-affected areas of the 2011 Great East Japan Earthquake and tsunami in Tōhoku had higher cohabitation rates across the generations within families, in which the main food source was from family-run fishing and farming businesses. No data on this change are yet available, but the media have reported that many families now live separately, whereas before the disaster they lived in three- or even four-generation households. There are a number of possible explanations, one being that the young and middle-aged might have moved to the city in the search of jobs and cash income. Moreover, children who lost their schools might have been moved to another town to have access to a school; while older people might have been left alone in their original houses, or in evacuation centres.

After a disaster, survivors are especially vulnerable, and life at evacuation centres and temporary shelters following a disaster raises challenges for all the main aspects of human security: 'freedom from fear', 'freedom from want', as well as 'freedom to live in dignity'. Life in evacuation centres may be temporary but such situations challenge the dignity of survivors. They often lack privacy. This may be more of a challenge for women. Losing their jobs also challenges the dignity of survivors. Immediately afterwards they are more concerned with surviving day to day, but over time they become more concerned with what they have lost because of the disaster. In this sense, both 'fear' and want' are present but shift from the emergency period over the mid and longer term. Survivors soon fear for what will happen to them and how they will be able to make a living when they have lost everything. And in the case of Fukushima there is also the fear that comes from the possible health effects of the nuclear radiation.

Whether an evacuation centre community works well or not is a community security issue. Health care and medical support at evacuation centres is positioned at the forefront of efforts to improve inhumane environmental conditions. Common basic needs such as privacy, hygiene and nutrition are often in scarce supply. Many survivors were located in large facilities with no privacy, such as crowded school gyms. Emergency toilets were often built outside the buildings, in open-air places. It was also cold for a prolonged period of time. Subsequently, many people, especially older people, restricted their eating and drinking habits because they did not want to wake up in the middle of night to walk to the uncomfortable toilets located outside in the cold. This caused many to suffer from dehydration and malnutrition at the time of the 1995 earthquake in Kobe (Tanida 1996; Otani 2010, p. 21). A similar situation occurred following the 2011 disaster at earthquake evacuation centres in East Japan.

There were also major health challenges following the 2011 earthquake and tsunami. As more than 30 per cent of evacuees were aged over 65, there was a high demand for routine medications used to treat chronic conditions such as hypertension, diabetes and heart disease. The major challenge for health providers, such as doctors, was the identification of the medicines which patients had been taking prior to the disaster. Often older patients did not remember the names of the specific medications they had been prescribed. Their pharmacy notebook records, if they had any, were commonly lost due to the tsunami. Pharmacists and public health nurses played a crucial role in the identification and selection of their medications and alternatives.

In order to address food insecurity at the evacuation centres scattered across vast areas of East Japan, food and drinking water were provided as emergency relief. These supplies often included *Onigiri* (rice balls) and bread, as these food stuffs are easier to deliver under such emergency circumstances. After previous disasters in Japan (Kobe and Niigata), people had to endure such food for the first few days, but soon then started to receive *obento* (lunch boxes), which contained a greater variety of food. However, in the case of 2011 Tōhoku, such food did not reach victims for days and sometimes weeks, due to geographical difficulties. This delay led to increased numbers of people reporting health problems caused by an unbalanced diet, which lacked vegetables and fruit. One NGO even chartered a helicopter to airdrop food supplies (vegetables and fruit) to evacuees in remote areas that could not be reached by car. In this situation a policy designed to address the threat of food insecurity led to the creation of another threat – that of health insecurity. This situation is an example of the way different human security threats can overlap and interrelate.

Food insecurity leading to health problems was also evident in issues related to the oral health of disaster victims after the 2011 earthquake and tsunami. Mental stress can trigger problems with the immune system, which can worsen the risk of developing cavities and decayed teeth. A dentist at Tōhoku University visiting temporary housing in Miyagi said that the prevalence of cavities seemed to have increased because of the changes in the diets of victims following the disaster. The bread provided as emergency relief tended to be sweet, as it was intended to last longer. This only worsened the oral health of victims, which was already difficult to maintain due to a shortage of toothbrushes and toothpaste after the disaster. Sunster, a company which produces oral health goods in Osaka, having experienced the 1995 Great Hanshin-Awaji Earthquake, took out a large oral hygiene advertisement in *Kahoku Shinpō* (the local paper in Tōhoku). The advertisement illustrated how evacuees could maintain oral health without toothbrushes and toothpaste. This is an example of corporate social responsibility (CSR), and indicates how companies too can play a role in contributing to the human security of disaster victims.

Another major health security issue is the lack of physical activity in the lives of evacuees. When staying at an evacuation centre all day long and without the opportunity to perform their work and daily chores, older people started to suffer from 'economy class syndrome' (deep vein thrombosis – blood clots often caused by long-distance airline travel). This can obviously be life threatening. The proportion of those who reported economy class syndrome was found to be higher among the elderly survivors of the 2011 Great East Japan Earthquake and tsunami than among the other victims. According to a survey conducted by Ishinomaki City in Miyagi, and the Ishinomaki Red Cross Hospital, from August 2011 to January 2012 at 18 compounds of temporary shelters, economy class syndrome was found in 43 out of 498 older residents, i.e. 8.6 per cent of the older population (*Yomiuri Shimbun* 2012). Some areas reported an even higher proportion. A doctor from Niigata visited three evacuation centres in Miyagi and found 11 elderly people out of 39 (i.e. 28 per cent) suffering from economy class syndrome, two months after the disaster (Kyodo 2011). Responding to this problem, medical teams visited evacuation centres to check the blood circulation of older people, especially in their legs. Public health nurses who visited older people who were immobile at evacuation centres provided massages to improve blood circulation and distributed pairs of pressured stockings. This was seen as a good way to keep the feet of the elderly warm in the cold weather and also to improve blood circulation.

While the 2011 earthquake occurred in the Japanese spring, temperatures for that particular March were below average. It remained cold for longer than usual and, to make matters worse, an unusual amount of snow started to fall in the disaster-affected areas, which made difficult conditions even more challenging. Despite the prolonged winter in 2011, the summer came earlier than expected, which compounded the challenges. Summer weather arrived in June, and the rainy season that followed was two weeks shorter than usual, meaning a longer and hotter summer. Throughout the summer, deaths of older people due to heat stroke were reported all across Japan, and not only in the areas directly affected by the disaster. Although it got much hotter earlier than usual, people were reluctant to use air-conditioning at a time when the government and electric power companies were asking people to save electric power, as power shortages were expected during the hot summer. Older people are known to be vulnerable to heat waves, not only in Japan but across the world. For example, in the 1995 heat wave in Chicago and the 2003 heat wave in Europe, older people tended not to feel the heat and did not notice they were dehydrated until it was too late.

Issues related to seasonal changes and weather fall within the category of environmental security. These concerns are closely connected to economic security, in particular energy security, not only for individual people going about their daily lives, but also for Japanese industry. These environmental security issues are also related to health security issues, as extreme weather – heat or cold – can impact on people's health, especially that of the elderly. Given the effect of the 2011 Great East Japan Earthquake and tsunami on the Fukushima nuclear disaster, nuclear power plants all over Japan were shut down. This meant there was less generation capacity to provide electricity until alternative power plants were built to meet increased demand. Some temporary shelters were given assistance to install heat-shut covers on their roofs, to shield them from excess heat during the summer. This option, however, is not considered to be as effective as electric air-conditioning. In the spring of 2012 there was much discussion surrounding the issue of how to cope with anticipated restrictions on electricity supply during the peak season in that coming summer. Media coverage addressed this discussion, but few practical energy-saving measures were proposed. As of 2013, a rotation plan was being considered to ration electricity, but is likely to have negative impacts on power-intensive industries, which require a constant and stable supply of power to operate. There was a growing concern for older people, not only for those affected by the disaster, but also all over Japan, who must cope in extreme heat with potentially limited power supplies.

Psychological effects

Mental health is an important component of health security, but it is also closely connected to community security. The psychological effects of disasters are a significant factor when examining post-disaster community security. Barusch (2011) argues that, while physical ageing and social indifference increase the physical vulnerability of elders, they may be more resilient than young adults to the psychological effects of disaster. Elderly disaster victims report less anxiety, show less stress, are less anxious and are less likely to suffer from long-term mental health problems than younger victims (Ngo 2001). However, these findings are qualified with the observation that older adults manifest their emotional reactions in ways that are less easy to detect. Otani (2010, pp. 5–27) found that research on the impact of disaster on mental health in old age has produced conflicting results. The elderly may be best prepared for disaster because of their previous life experiences, or they may be more vulnerable than younger age groups because of their frail health, strong emotional attachments to long cherished property and mementoes that have been lost, lower adaptability than in their younger days and/or because they tend not to discuss their problems spontaneously unless they are questioned specifically (Tanida 1996; Gerrity and Flynn 1997).

Different studies, however, have reached opposite conclusions. Meanwhile, Robertson and Bell (1976) concluded that the old and poor were reluctant to use available resources, and the shock of a natural disaster seemed to last longer among elderly individuals.

The findings of research on post-traumatic symptoms by Kato et al. after the Great Hanshin-Awaji Earthquake concluded that the elderly were better protected from the stresses of disaster when compared with younger people. During the first assessment, in the third week, subjects who were younger than 60 years old and those older than 60 years old experienced sleep disturbances, depression, hypersensitivity and irritability. During the second assessment, in the eighth week, the percentage of younger subjects experiencing symptoms did not decrease, while elderly subjects showed a significant decrease in eight out of ten symptoms (Kato et al. 1996). Kato et al. considered three explanations for their findings. First, the younger evacuees may have experienced greater psychological stress in reconstructing their lives and those of their families. This may include finding new jobs. In comparison, older people in the shelters are often living on their pensions and some personal savings, or as recipients of welfare, which meant that finding a new job was not a problem for them. Secondly, the elderly might have established better social networks in the shelters than the younger survivors because they had lived in the pre-earthquake local community for longer. The third explanation relates to previous disaster experiences. Kato et al. also discussed other limitations of their study, such as the interpretation of the data and the absence of controls. Symptoms such as sleep disturbances and irritability do not necessarily predict the degree of psychological impairment.

Knight et al. (2000) have pointed out that mental ill-health following a disaster cannot be solely attributed to the disaster. For example, in a study of the mental health of older people after the 1994 Northridge Earthquake in Los Angeles, USA, prior earthquake experience was related to lower post-earthquake depression scores. The strongest influence on post-disaster mental health was pre-disaster mental health (Knight et al. 2000). In Kobe, it is conceivable that older earthquake survivors in temporary shelters were already suffering from depression related to loneliness and isolation, even before the earthquake.

Mental health effects can be divided into three categories: (1) short-term mental disorders due to the shock of an earthquake; (2) the development of latent diseases triggered by the shock and stress of an earthquake; and (3) those symptoms caused purely by the earthquake, including post-traumatic stress disorder (PTSD). In the second category, most frequently reported are dementia cases triggered by mental stress and environmental change, as well as alcohol-dependency syndrome (Otani 2010, p. 25). Dementia is reportedly triggered or worsened for many older people after earthquake events. This was an issue in the 1995 Great Hanshin-Awaji Earthquake (Otani 2010) and has been a concern for older people and the community following the 2011 East Japan earthquake and tsunami. Sudden changes in living environment have been known to trigger or worsen dementia in elderly persons. As the post-disaster temporary evacuation environment is an environment in flux, elderly persons with pre-existing dementia can experience further diminishment of their mental faculties. Providing community security, closer to that available pre-disaster, is key for older people to prevent or lessen dementia. A functioning community where people know each other and have good relationships, rather than being surrounded by strangers, is key for taking care of demented older people. This is an important reason for prioritizing strategies that incorporate community security by keeping communities together in the same shelters and temporary accommodation, if possible.

Community security for an ageing society

Harasawa et al. (2012) report that one of the most tragic aspects of the 3/11 disaster in Fukushima was the break-up of communities and families due to fears of radiation exposure. The population of Minamisoma City, Fukushima Prefecture, saw major changes in population, as a direct result of concerns about radiation. Prior to the nuclear disaster the population of the city was 72,000. Immediately afterwards this figure dropped to 10,000, and had only recovered to 43,000 one year later. Of note, the proportion of those aged 65 years or older has increased from 25.9 per cent to 32.2 per cent. Many young families moved out of the city, which resulted in a sudden increase in the number of frail elderly people living alone in temporary housing, some of whom subsequently died alone in their homes. Forming a new community, or restarting a community that was destroyed by disaster, and building up social capital is not an easy task, especially in a community with an ageing population. Achieving community security becomes more challenging in an ageing society.

What was learnt from Kobe was that special attention should be given to the elderly in terms of both allocation and design of temporary housing. Gaps in floor levels may cause problems for elderly and disabled residents. The shortage of common space, green areas and living facilities could affect the mental health of residents. The shortage of common areas was a factor in the difficulties experienced generating a community (Otani 2010, p. 97). Previous experience has taught that we need to pay special attention to older survivors and that they should not be left in isolation (Otani 2010, p. 132). From a study on older people in the 1999 Taiwan Chi-Chi Earthquake, among the significant factors associated with a lessening of depressive symptoms were social support from the extended family and neighbours, and social participation (Watanabe et al. 2004).

Intervention to promote increased social networks and social participation within victims' new temporary communities is vital, especially for older people. The age-specific approach stands in contrast to guidelines for best practice developed by HelpAge International (Cunningham 2011). Based on the experiences of older adults in disasters around the world, HelpAge recommends against the establishment of separate response plans for older people. Instead, it is suggested that emergency plans for the general population should take into account the needs and vulnerabilities of older people.

Japanese society is very concerned about preventing *kodokushi* (dying alone). This topic was highlighted in a report in the early 1990s but it became a headline-grabbing concern at temporary shelters and public housing in the years following the 1995 Great Hanshin-Awaji Earthquake (Otani 2010, p. 39). In the Kobe case, *kodokushi* was commonly reported in men in their fifties and women aged 70 and above. There was a clear gender and age difference in this pattern. These were also the same gender and age groups that were most vulnerable in the former Soviet Union after the political and economic transitions of the early 1990s (Otani 2010, pp. 81–2). To prevent isolation and dying alone, we need to consider post-disaster community security. During rebuilding it is necessary not only to construct the required number of temporary houses, but also to design common spaces for people to meet. This would most certainly engender a sense of community, and would allow residents to become 'neighbours' in the new community. This would also lead to an increase in opportunities for improved physical activity and diet.

After the Kobe experience, it was recognized that it was important not only to build structures which would be earthquake resilient, but also in a way that took into account the needs of the elderly and the disabled. Observation of the Kobe case shows that, after one year, there was a change in the profile of those who remained in temporary shelters, and that the kind of support and assistance needed changed. After the earthquake, people from all strata of society lost their houses and moved to evacuation centres – often to school gyms, and then to temporary shelters. Those who could 'stand on their own', who had more money or who were eligible for housing loans, were more likely to move from temporary shelters and begin to move on with their post-disaster lives. It was those who remained that needed more help. Many of these were marginalized people, including those living on welfare, with alcohol dependency or those with no family. As a result, in Kobe temporary shelters that were built to last for one year were occupied for more than three years. This experience suggests that when building temporary housing special attention needs to be paid to the most vulnerable parts of that community.

Gender and post-disaster community for the elderly

Looking at post-disaster community security through a gender lens is also important, when considering the elderly. From a human security perspective, gender-based violence is not only an issue of community security but also personal security. Violence against women in post-disaster communities has been reported to be a serious issue in many disasters – Hurricane Katrina in New Orleans, the

Haiti earthquake (Davoren 2012) and Sri Lanka tsunami are but a few examples (Fisher 2010). In the case of the 2011 earthquake and tsunami, Saito (2012) found that cases of sexual and domestic violence against women were reported. Research conducted by one women's group highlighted 14 cases, while another women's organization reported that about 15 per cent of the 324 calls to consultation services, between 29 March and 23 July 2011, were to do with violence against women (Osawa et al. 2011; Sendai Gender Equality Opportunity Foundation 2011). Saito observed that young women, both evacuees and volunteers working in the evacuation centres, suffered from sexual harassment. It was also reported that an old man was beaten to death at a temporary shelter in Ishinomaki in August 2011. When a place is so unsafe that a man can be treated so violently, what chance would there be for a woman living alone there? Gender-based violence is not limited to younger women, older women also suffer. After the Kobe earthquake there was a case of a woman in her seventies being subject to sexual abuse by a man in his sixties at a temporary shelter.

The 2011 Great East Japan Earthquake and tsunami indicated again a failure on the part of government to integrate a gender perspective into emergency planning and response. Saito (2012) identified, from her experience working in Tōhoku, the reinforcement of traditional gender roles, and a lack of privacy (including for seemingly simple issues like the drying of women's underwear). It can be argued that these problems are reflective of ongoing gender inequality issues in Japanese society (Saito 2012). An ageing population gives rise to a higher proportion of females. In Japan, as in other countries, the number of older females exceeds the number of older males. The position of women is critical in post-disaster communities – both in the initial emergency period and in longer-term post-disaster reconstruction (Otani 2010). What this suggests is that it is necessary to address gender-related issues in order to promote both personal security and community security.

Conclusion

Community security, understood as the forming of a functioning community, is key to recovery and resilience at all stages of the post-disaster process. This focus on community should begin immediately following the emergency, first in the evacuation centres, then in the temporary shelter communities in the mid-term and, finally, in the reconstruction of permanent housing. As has been argued, specific attention must be paid to addressing the specific challenges that come from an ageing society. For older people, it is not simply a case of 'being taken care of', but being included in the fabric of the community. Paying more careful attention to community security issues can have a positive influence on other aspects of human security – including personal and heath security. In this way, addressing issues related to the elderly can have empowering consequences.

The way elderly people respond to disasters is just as diverse as any other age group. As such, it is necessary to look beyond the stereotypes and headlines. While acknowledging this, previous experience indicates the need to pay special attention to older survivors, taking extra care and attention not to leave them in isolation, especially those lacking extended family support. *Kodokushi* (dying alone) became a major issue following Kobe as the media covered the plight of lonely elderly people living in isolation, without the mental or physical resources to cope with such a traumatic disruption of their lives.

To prevent isolation, it is critically important to design common spaces that facilitate daily interaction. It is not just a case of the numbers of units or of their seismic strength, but also the creation of aesthetically attractive compounds with barrier-free design for the elderly and the disabled. These initiatives are designed to address community insecurity by facilitating accumulation of social capital and fostering the sustainability of social networks. This also has positive consequences for other human security issues, such as health security. When people are able to share in cooking and eat together, as in more traditional rural lifestyles, it can have beneficial consequences for both community security and health security.

This chapter has focused on older people and community security within post-disaster communities with ageing populations. The chapter also introduced some examples of the unintended impacts of policies, indicating how addressing one threat could lead to greater insecurity in another area. When looking at the way elderly people are impacted by disasters, it can be seen that the different threats to human security are closely interrelated, and careful attention needs to be paid to the way they connect and overlap. As this chapter has shown, post-disaster life raises serious challenges for all the major components of human security – 'freedom from fear', 'freedom from want' and 'freedom to live in dignity'. A human security approach must show an awareness of the different strengths and weaknesses of different age groups, and through this, strengthen levels of community security.

References

Aldrich, D. 2012, *Building Resilience: Social Capital in Post-Disaster Recovery*, Chicago, IL: University of Chicago Press.

Barusch, A. S. 2011, 'Disaster, vulnerability, and older adults: Toward a social work response', *Journal of Gerontological Social Work*, 54, pp. 347–50.

Cabinet Office 2011, Government of Japan, Aging Society White Paper, http://www8.cao.go.jp/kourei/whitepaper/w-2011/zenbun/pdf/1s1s 2.pdf, viewed Dec. 2013.

Chan, E. Y. Y. 2008, 'The untold stories of the Sichuan earthquake', *The Lancet*, 372, pp. 359–62.

Cunningham, R. 2011, 'Ageing in the media: Japan earthquake and tsunami', *HelpAge International*, 24 March, http://www.helpage.org/blogs/rosaleen-cunningham-21/ageing-in-the-media-japan-earthquake-and-tsunami-239, viewed Dec. 2013.

Davoren, S. J. 2012, 'Helping international non-government organisations (INGOs) to include a focus on gender-based violence during the emergency phase: Lessons learned from Haiti 2010–2011', *Gender and Development*, 20(2), pp. 281–94.

Enarson, E. 2014, 'Human security and disasters: what a gender lens offers', in C. Hobson, P. Bacon, and R. Cameron (eds), *Human Security and Natural Disasters*, London: Routledge, pp. 37–56.

Fisher, S. 2010, 'Violence against women and natural disasters: Findings from post-tsunami Sri Lanka', *Violence Against Women*, 16, pp. 902–18.

Gerrity, E. T. and Flynn, B. W. 1997, 'Mental health consequences of disasters', in E. K. Noji (ed.), *The Public Health Consequences of Disasters*, New York and Oxford: Oxford University Press, pp. 101–21.

Harasawa, K., Tanimoto, T., Kami, M., Oikawa, T., Kanazawa, Y. and Komatsu, H. 2012, 'Health problems in the temporary housing in Fukushima', *The Lancet*, 379, pp. 2240–1.

Kato, H., Asukai, N., Miyake, Y., Minakawa, K. and Nishiyama, A. 1996, 'Post-traumatic symptoms among younger and elderly evacuees in the early stages following the 1995 Hanshin-Awaji earthquake in Japan', *Acta Psychiatrica Scandinavica*, 93(6), pp. 477–81.

Keeling, S. 2011, 'How are older people in Christchurch doing?' New Zealand Association of Gerontology, June, 11, http://www.iagg.info/data/NEW_ZEALAND_Newsletter_May2011_0.pdf, viewed 16 March 2014.

Kingston, J., ed. 2012, *Natural Disaster and Nuclear Crisis in Japan: Response and Recovery After Japan's 3/11*, London: Routledge.

Klinenburg, E. 2002, *Heat Wave: A Social Autopsy of Disaster in Chicago*, Chicago, IL: University of Chicago Press.

Knight, B. G., Gatz, M., Heller, K. and Bengtson, V. L. 2000, 'Age and emotional response to the Northridge Earthquake: A longitudinal analysis', *Psychology and Ageing*, 15(4), pp. 627–34.

Kyodo 2011, 'Hinan jumin no sanwarijaku ni kessen – miyagi de ishi ga kanikensa' (Blood clots observed among nearly 30% of evacuees – Doctors conducted screening tests in Miyagi Prefecture), 24 March, http://www.47news.jp/feature/kyodo/news04/2011/03/post-898.html, viewed Dec. 2013.

Low, W. Y. and Binnes, C. 2011, 'Disasters and public health concerns, *Asia-Pacific Journal of Public Health*, 23(3), pp. 277–9.

Morrow, B. H. 1999, 'Identifying and mapping community vulnerability', *Disasters*, 23(1), pp. 1–18.

Ngo, E. B. 2001, 'When disasters and age collide: Reviewing vulnerability of the elderly', *Natural Hazards Review*, 2, pp. 80–9.

Osawa, M., Domoto, A. and Yamaji, K., with Minagawa, M., eds. 2011, 'Saigai Fukko to Danjo Kyodo Sankaku (Disaster Reconstruction and Gender Equality)', *6/11 Symposium*, ISS-GCOE Research Series No. 4, ISS Research Series, 46, Tokyo: Institute of Social Science, University of Tokyo.

Otani, J. 2009, 'Issues in modern China highlighted by the 2008 Sichuan earthquake with reference to the 1995 Great Hanshin-Awaji earthquake and the 2005 West Off Fukuoka earthquake' [Japanese with English abstract]. *Bulletin of Kyushu University Asia Centre*, 3, pp. 23–37.

Otani, J. 2010, *Older People in Natural Disasters*, Kyoto and Melbourne: Kyoto University Press and Trans-Pacific Press.

Otani, J. 2011, 'HIV/AIDS, SARS and natural disasters', in S. Amako (ed.), *Non-Traditional Security in Asia 1: An Overview*, Keiso Shobo.[Japanese], 49–78.

Otani, J. 2012, 'Ageing society, health issues and disaster: Assessing 3/11', in J. Kingston (ed.), *Natural Disaster and Nuclear Crisis in Japan: Response and Recovery After Japan's 3/11*, London: Routledge, pp. 237–54.

Phifer, J. F., Kaniasty, K. Z. and Norris, F. H. 1988, 'The impact of natural disaster on the health of older adults: A multiwave prospective study', *Journal of Health and Social Behaviour*, 29(1), pp. 65–78.

Robertson, L. and Bell, B. 1976, *Planning for the Elderly in Natural Disaster*, Omaha, NE: University of Nebraska.

Saito, F. 2012, 'Women and the 2011 East Japan disaster', *Gender and Development*, 20(2), pp. 265–79.

Sendai Gender Equality Opportunity Foundation 2011, *Bulletin*, 27, Nov., Sendai: Sendai Gender Equality Opportunity Foundation.

Seplaki, C. L., Goldman, N., Weinstein, M. and Lin, Y. H. 2006, 'Before and after the 1999 Chi-Chi earthquake: Traumatic events and depressive symptoms in an older population', *Social Science and Medicine*, 62, pp. 3121–32.

Tanida, N. 1996, 'What happened to elderly people in the Great Hanshin Earthquake?', *British Medical Journal*, 313(7065), pp. 1133–5.

Varley, A. 1994, *Disasters, Development and Environment*, Chichester: John Wiley and Sons

Wang, M. 2011, 'Towards "non-traditional security" network construction: Roles of NGO/NPOs', in S. Amako (ed.), *Non-Traditional Security in Asia 2: China in the Region*, Keiso Shobo [Japanese], pp. 13–37.

Watanabe, C., Okumura, J., Chiu, T. Y. and Wakai, S. 2004, 'Social support and depressive symptoms among displaced older adults following the 1999 Taiwan Earthquake', *Journal of Traumatic Stress*, 17(1), pp. 63–7.

Yamamoto, T., Kato, M. and Shirabe, S. 2011, 'Life, health, and community in a tsunami-affected town', *The Lancet*, 23 July, pp. 378, 318.

Yomiuri Shimbun 2012, 'Ekonomii kurasu shoukougun – kasetsu no koreisha ichiwari ga hasshou' (Economy class syndrome observed among 10% of older people living in temporary housing), 3 Feb.,http://www.yomiuri.co.jp/feature/eq2011/information/20120203-OYT8T00179.htm, viewed Dec. 2013.

9 Post-disaster recovery and the cultural dimension of human security

Akiko Fukushima

Introduction

The concept of human security was originally introduced in the UNDP's seminal 1994 *Human Development Report*. A decade later, UN member states officially acknowledged the term in the 2005 World Summit Outcome document, as follows:

> We stress the right of people to live in freedom and dignity, free from poverty and despair. We recognise that all individuals, in particular vulnerable people, are entitled to freedom from fear and freedom from want, with an equal opportunity to enjoy all their rights and fully develop their human potential. To this end, we commit ourselves to discussing and defining the notion of human security in the General Assembly.
>
> (UNGA 2005, paragraph 143)

As the above paragraph indicates, member states could not agree precisely on how to identify the characteristics of human security, and simply committed themselves to discussing the matter further. As a result, human security was discussed at the UN, and the outcomes of these discussions were noted in General Assembly resolutions and the Secretary-General's subsequent reports on human security.

In 2010 the UN Secretary-General issued a report on human security which adopted a broad definition of human security that included 'freedom from fear' and 'freedom from want', as well as 'freedom to live in dignity'. The Secretary-General, mindful of the concerns of some member states, explicitly discounted the use of force, despite the fact that 'freedom from fear' was included in the definition. The report also asked states to mainstream the approach in UN activities (UNSG 2010).

On 5 April 2012, the Secretary-General submitted a second report, which suggested offering a possible 'common understanding on the notion of human security' (UNSG 2012), as a way of allowing the term to become established at the heart of UN discourse. The second report also emphasized the fact that human security is a practical policy framework for addressing widespread and cross-cutting threats. It allowed actors to institute protection strategies and empowerment measures, and to take 'a comprehensive and integrated approach' (UNSG 2012). This certainly also applies to disaster situations, which are explicitly included in threats to human security 'as evidenced by the recent earthquake and tsunami in eastern Japan' (UNSG 2012). In advocating the mainstreaming of human security, the report cited four areas of UN activity in which a human security approach can be useful, including the area of climate change and climate-related hazard events (UNSG 2012). The report therefore leaves no doubt that Japan's triple disaster and natural disasters in general are covered by the notion of human security and threats thereto.

If human security is to become a mainstream principle, certain elements must be more fully developed before the international community can adopt the term as a guiding concept. Below I focus on two aspects of what is missing. First, since its inception, the term human security has been used with reference to issues related to foreign, not domestic, policy. Yet, in light of the elements covered by the term, I believe that the concept should be used in the context of both international *and* national policies, regardless of whether countries are developing or developed, and also that human security is highly relevant to issues such as poverty, health, the environment and natural disasters. This is something Japan discovered following the triple disaster of March 2011: previously it had strongly championed the human security concept, but only as part of its foreign policy.

Second, it should not be overlooked that culture has been identified as an integral element of human security. This has been noted in many reports and recommendations, such as the 2003 Commission on Human Security report (CHS). The commission's definition of human security is 'to protect the vital core of all human lives in ways that enhance human freedoms and human fulfilment', adding that this 'means creating political, social, environmental, economic, military and cultural systems that together give people the building blocks of survival, livelihood and dignity' (CHS 2003, p. 4). However, the report does not expand on the content of cultural systems required for human security other than in the context of education, in connection with which it states that schools and their teachers can teach mutual respect and solidarity (CHS 2003, pp. 122–3). Thus, while culture has been recognized as an important dimension of human security, its role has not been sufficiently emphasized.

Below, I attempt to address these gaps in the human security literature by looking at some of the cultural issues which are relevant in the context of natural disasters. Since natural disasters involve not only natural phenomena, but also human agency, a comprehensive approach to human security is called for throughout the three phases of prevention, response and recovery. I argue that cultural activities can play a highly significant role in the psychological rehabilitation of people and communities in the late-response and recovery phases following natural disasters. In presenting this argument, I draw on my research on the role of cultural activities conducted as part of peace-building efforts, which were undertaken against a background of violent conflict, to garner human security for people in conflict-ridden areas. I then examine events surrounding the March 2011 Great East Japan Earthquake, the tsunami and the Fukushima nuclear crisis through a culturally-informed human security lens. The chapter ends with some conclusions and recommendations.

Natural disasters and human security

The consequences of natural disasters, such as earthquakes, tsunamis, floods and hurricanes, are difficult to manage, and recovery and rebuilding are not just about physical reconstruction. Faced with death, injury, destroyed infrastructure and debris, survivors who have lost family members and friends, belongings and even their homes, experience a profound sense of loss that causes long-lasting trauma. Some survivors fear another disaster, some suffer stress from having to live in shelters and then temporary accommodation, until towns and housing have been rebuilt and they can return to the places they formerly called home.

In assisting disaster-hit areas, priority is often initially given to (1) supplying basic daily needs and the removal of debris and mud. (2) The reconstruction of infrastructure and housing is then undertaken; before (3) the physical health of survivors is addressed. Psychosocial aspects – the 'soft' side – of reconstruction are considered of secondary importance. Certainly, consequence management – the 'hard' side – has top priority, as survivors must have the means to live in the days after a disaster. But can the 'soft' side of recovery be dismissed as marginal? Unless people who are afraid of a follow-up earthquake and major aftershocks can find a way of letting go of their fear, they cannot turn their attention to reconstruction. It is the people who matter in reconstruction for, unless survivors regain the strength to work for the future, they will be no more than passive witnesses to reconstruction, thereby impeding the rejuvenation of the area. Disaster consequence management plans must therefore include the 'soft' side of reconstruction and take a people-centred approach if the area is to experience robust recovery. This also matches closely with the 'freedom from fear' component of human security mentioned above.

The 1994 Human Development Report (HDR) clearly sets out a people-centred approach. It also recognizes threats to environmental and community security in its discussion of the seven dimensions of human security. With regard to environmental security, the report argues that environmental degradation affects both developing and industrial countries, and includes natural disasters among the threats which are being experienced with growing frequency and severity. In terms of community security, the report argues that people derive security from the group of which they are a member, and with which they share cultural identity and a set of values. Both factors are key elements of human security in disaster-hit areas (UNDP 1994, pp. 31–2). Thus, the seminal HDR report incorporates the relevant elements of a post-disaster context in the notion of human security that it espouses.

This leads one to ask whether cultural activities have a role in the broader context of post-natural disaster consequence management. In considering the possible value of such activities for recovery from natural disasters, some insight may be gleaned from the role of cultural activities in post-conflict peace-building. In fact, some peace-builders have encountered natural disasters during their peace-building activities, and applied their techniques to disaster victims, as I discuss later in the chapter. Moreover, many peace-building organizations and individuals in conflict situations overseas came to assist people in the Tōhoku area after the 11 March 2011 triple disaster, and found that they could apply their peace-related techniques and know-how in post-disaster situations.

Peace-building and cultural activities

In the wake of the growing number of intra-state conflicts of the post-Cold War era, the international community rallied to assist in the post-conflict reconstruction of war-torn countries, many of which had severely weakened governance structures. Many countries emerging from prolonged periods of conflict simply do not possess the requisite capacity to recover and thus rely, partly, on the peace-building assistance provided by the international community.

While peace-building is best initiated before the outbreak of conflict, when there remains a chance of preventing armed clashes, the main stage for such activity unfolds following the conclusion of peace accords. This calls for a comprehensive, multi-faceted set of efforts, encompassed under the general rubric of peace-building (political, economic and security). Nevertheless, Paul Collier and colleagues point out that, in the case of civil wars, 44 per cent of conflicts recur within five years of peace accords having been reached (Collier et al. 2003, pp. 83–4). This high percentage suggests something is lacking in current peace-building efforts. One missing element might be addressing the cultural dimension of recovery, which observation led the author to consider the role of culture in peace-building (Fukushima 2012, pp. 17–20).

Helping populations

In comprehensive, multi-faceted peace-building, cultural activities have served wide-ranging functions, which may provide insight into their possible value in helping people recover from natural disasters. Some of the functions are outlined below:

1 As catalysts and facilitators for mutual understanding and accommodation among former enemies, leading to reconciliation. This is necessary in peace-building, since the parties must live together in the same or neighbouring communities after intra-state conflicts, and the recurrence of conflict must be prevented. In contrast, following inter-state wars former enemies live separately, in areas demarcated by agreed national borders.
2 As a means of turning the post-conflict despair of local populations to hope for the future.
3 As a means of caring for and healing the psyche of survivors, so that they can overcome their conflict-related trauma.
4 As a way of assisting local populations to rebuild their self-respect and the pride lost during times of conflict, enabling them to transcend hatred and fear of former enemies, and so sublimate identities deformed or damaged by conflict-related experiences.
5 As a means of empowering local people to become actively involved in peace-building.

Among these functions, caring for and addressing the trauma of survivors may have the greatest relevance in a post-disaster context. In conflicts, people suffer an acute sense of loss when family members and friends are killed and enemies force them to leave their homes and flee without their possessions, while their houses are burnt to the ground. In refugee camps, people must live in an environment replete with emotional and social disturbance, and experience confusion, fear, anxiety and loneliness. Stress levels soar as people find their possessions gone and must live in groups among strangers. Forced reliance on assistance from the outside also affects people's self-respect, often depriving them of hope and prospects for the future. These experiences are shared by survivors of both conflicts and disasters, because the recovery and reconstruction phases are structurally similar, as is noted in the Hyogo Framework for Action, set up in 2005 under the auspices of the United Nations International Strategy for Disaster Reduction (UNISDR).

Healing the psyche

According to a 1997 survey conducted in Sarajevo by the United Nations Children's Fund (UNICEF), one-third of the 422 children evaluated were severely traumatized, while 15 per cent of the entire population showed some adverse psychological symptoms resulting from their war experiences. Since the conflict ended, there has been an increase in the number of children and young people suffering from post-traumatic stress disorder (PTSD), with cases of suicide being noted.

Because recurring conflicts are often caused by psychologically scarred individuals, it is important that such wounds are attended to, particularly since there are a number of conventional clinical methods for delivering psychological care. Moreover, to supplement conventional medical care, cultural activities and participatory events, featuring art and theatre, have been shown to produce positive outcomes (Fukushima 2012, pp. 151–82). Drawing, which reveals the condition of an individual's innermost feelings and the state of the soul, can serve as a vehicle for healing.

Painting workshops have allowed children to open up enough to express emotions, as has been the case, for example, with children in Bosnia-Herzegovina. When asked to paint something after the conflict, many could draw nothing at all, while some who found it in themselves to produce a drawing used only the colour black. It took many children a couple of years before they could use colours such as pink, yellow and red for flowers, and green for the leaves on trees. But this healing process put some sparkle back in the children's eyes, as the author witnessed during a research visit to Bosnia-Herzegovina in 1998 (Fukushima 2012).

Bi-directional theatre workshops are another method to care for psychological wounds. Although people find it difficult to convey their emotions, composing and taking part in skits has proved to be an excellent tool. Participants generally say that they are finally able to express deeply buried feelings; they have discovered that their enemies went through similar horrible experiences; and they have come to terms with their conflict experiences (Fukushima 2012, pp. 159–73).

Rebuilding self-respect

Also important in healing psychological wounds is rebuilding a sense of self-respect and pride, which is often lost during wartime. One way of doing this is by restoring people's cultural heritage, as is being done through the work in progress at Angkor Wat in Cambodia. The same was attempted through the restoration (completed in 2004) of the Mostar Bridge in Bosnia-Herzegovina.[1] Such renovation work allows local people to recognize the value of their heritage, to take pride in it and to receive revenue in the future from tourism, rather than always having to rely on international assistance.

Restoration of traditional cultural industries generates a similar effect, as has been seen in the Afghanistan city of Istalif, 50 km north of Kabul. Formerly, the city was a summer resort that boasted vineyards and Istalif pottery – brilliant blue, green and turquoise glazed ceramics in a style that dates back more than 300 years. The robust nature of the pottery makes it perfect for dishes and pots, which used to be sold by some 800 merchants at the Istalif bazaar.

Due to its strategic location, the city was the site of battles between Taliban government forces and the Northern Alliance. As crops, homes, pottery studios and kilns were destroyed, even the 300-year-old kiln fires were extinguished, causing equally old traditions and skills to die out. Although after the fall of the Taliban government in 2001, citizens who had fled to Pakistan and Iran during the war began to return and the kiln fires were rekindled, it was then left mainly up to village youths to restore the traditional technologies. This they could do only by trial and error.

In 2002, a Japanese group visited Istalif to assist in the revival of traditional pottery-making skills, and invited Istalif potters to tour Japanese pottery towns and learn about Japanese pottery technologies. During their time in Japan, the Istalif potters became acquainted with state-of-the-art electricity-powered pottery wheels and modern glazes. Inspired to restore the city's pottery, which they now sell domestically and export, young potters have become acutely aware of their heritage, and take pride in being Afghan and generating income, rather than relying solely on foreign aid. Clearly, culture has played a major role as an important part of a comprehensive and multi-faceted peace-building mechanism (Fukushima 2012, pp. 37–60). Many of these experiences could be extended when thinking about how people can recover from the similar experiences of stress and trauma they suffer after a natural disaster.

Disaster recovery and cultural activities

Some research has been carried out on the connection between natural disasters and the cultural dimension of recovery. For example, Elaine Enarson has suggested that disaster victims can manage to survive a difficult recovery period by telling their own stories of a particular disaster. This they do through popular culture, such as art, poetry, music, drama and fabric art. The activities help survivors come to terms with the shock and loss they have experienced. In the article, Enarson takes up the example of quilting, telling of how quilts represented the 1997 spring flooding experiences of residents along the Red River, in the twin cities of East Grand Forks, Minnesota, and Grand Forks, North Dakota.

Enarson notes how flood survivors, particularly women – who had tried and failed to save the city through weeks of intensive sandbagging – were forced to evacuate at night and returned weeks or even months later to muck out the debris and make decisions concerning rebuilding. It was these people who told their stories in quilts. Enarson examined two quilts, one made by university students, and the other by a guild of older women. The former represented how the flood took place, its root causes and the imbalance between man and nature. This quilt was displayed on the first anniversary of the flood, draped over an antique quilt frame that had been rescued from a flooded home. The second quilt comprised sections made by guild members, in which they expressed the impact of the flood on their lives. Many parts of the quilt used dark colours, while others embraced hope for the future with motifs in bright hues. Local quilters explained that working to try and contain the flood waters had left them completely exhausted, and that they had needed an alternative activity to give them a sense of hope and optimism (Enarson 2000). This case study suggests that cultural activities, here the making of traditional quilts, had helped local residents to come to terms with the flood by providing them with an outlet to express their experiences and emotions.

As this example and the previous examples from post-conflict scenarios illustrate, conflicts and disasters have many features in common when it comes to the impact they have on local populations. Affected populations benefit from artistic outlets for self-expression, in both post-conflict and post-disaster contexts. Given these similarities, I suggest that cultural activities – ranging through sport, music, theatre, traditional festivals and items of national heritage – may play an equally important part in both post-conflict rebuilding and post-disaster recovery. At such times, the psychological scars of survivors must be treated and healed; ways must be found to counter survivors' apathy and turn despair in the direction of hope; community ties must be revived when villages have been destroyed in combat, washed away by floods or reduced to embers by fire; and survivors must be empowered to rebuild their homes, towns and, ultimately, their lives.

As mentioned earlier, the members of several Japanese NGOs active in peace-building overseas, in both post-conflict and post-natural disaster situations, have found that their know-how and techniques can produce similar effects in assisting survivors who suffer from a sense of loss and often feelings of apathy. One such individual is Norihiko Kuwayama, a medic specializing in psychosomatic medicine, who hails from Miyagi Prefecture, Tōhoku. He is also a representative of the NGO Stage Earth, and has been active in assisting people in conflict-ridden areas, including Palestine and Timor Leste, where he has provided medical and psychosocial care for those who experienced conflict-related trauma. In Palestine, he has regularly visited Gaza since 2004, and has provided medical care as well as organized painting and clay-making classes for children aged between 12 and 15.

Following the 1999 turmoil in Timor Leste, Kuwayama helped a clinic in Dili when the full-time doctor was on leave, and organized a theatre company made up of Japanese volunteers who performed a skit titled 'Mango Tarō', which is an adaptation of the Japanese fairy tale 'Momotarō'. While the peach (*momo*) is a familiar fruit in Japan, the troupe changed the word to mango, because it is a fruit with which the people in Timor Leste are more familiar. The troupe's performances in villages drew a great deal of laughter, something that had not been heard during and immediately after the 1999 turmoil.

The doctor also applied his techniques to children who survived the 2003 earthquake in Java, Indonesia. While providing medical care to the victims of the earthquake, he at the same time organized a musical session with local children and composed a song with them which they all enjoyed. Kuwayama argues that cultural activities, such as theatre, music and sport, provide people with a means of releasing stress that otherwise they would have to suppress.

More recently, Kuwayama has similarly helped survivors of Japan's devastating 11 March 2011 earthquake. His clinic was hit by the earthquake and tsunami, yet he has been busy assisting local people, organizing projects in which children in Miyagi Prefecture create dioramas of local towns before and after the tsunami, as well as ones depicting how they imagine the areas might look in the future. He has also organized a crochet class for area housewives who produce for sale rainbow-coloured scrubbies and dolls depicting local guardian deities of children. Applying to his home region of Miyagi the know-how and techniques Kuwayama uses in conflict areas has helped both to lift people out of their despair, and to generate a sense of hope for the future.[2] These cultural activities of the NGO Stage Earth as part of response, recovery and reconstruction efforts following the earthquake, tsunami and Fukushima nuclear crisis of March 2011 provide an excellent illustration of the connections and similarities between peace-building and disaster recovery.

The Great East Japan Earthquake

The magnitude 9 earthquake occurred at 2:46pm (5:46am GMT) in Tōhoku, and particularly affected the prefectures of Miyagi, Fukushima and Iwate. Lasting six minutes, it was followed by tsunami waves that reached heights of up to 40.5 metres (in Miyako, Iwate Prefecture) and travelled up to 10 km inland (Sendai area), tearing asunder the landscape and wiping away many townships along Japan's northeastern Pacific coast. As a result, six reactors at the Tokyo Electric Power Company's Fukushima Dai-ichi nuclear power plant suffered crippling damage that led to radioactive contamination of the area adjacent to the plant and beyond, endangering the lives of a large sector of the population and sowing terror.

The Japanese have long faced major earthquake-related catastrophes and recognize the associated dangers. Yet, despite the massive Jōgan earthquake and tsunami that occurred in 869, during the Heian period (794–1185), leaving sand deposits several kilometres inland, the associated lessons have been ignored by many. Construction has flourished along the coastline, ignoring the hundreds of centuries-old stone markers and tablets warning later generations of the danger that tsunamis may pose. After the 2011 quake, people in Tōhoku recognized the importance of these old stones and are bearing in mind their message as they focus on reconstruction.

Assessing the costs

The losses incurred as a result of the triple disaster are huge. In terms of casualties, the earthquake and tsunami left 15,885 people dead and 2,823 missing (as of 10 April 2014, according to a National Police Agency of Japan Report in 2014). In monetary terms, the Cabinet Office expected the tally, for residents' tangible assets and regional infrastructure, to reach some ¥16.9 trillion (Reconstruction Design Council 2011, p. 7). Safety-related rumours continue to plague domestic sales and exports of fresh produce and manufactured products from the affected areas.

Japan's Reconstruction Agency reported that, as of the end of 2013, approximately 305,000 evacuees remained in temporary shelters and housing (Japan's Reconstruction Agency 2014). A 20-km exclusion zone was introduced around the nuclear plant and, while citizens were permitted to return to some areas, they have been prevented from returning to others, such as the town of Namie, where there is a strong possibility that the exclusion order may remain in place for the next several years. Not only have evacuees had to move to temporary shelters, but some have had to move several times, either for their safety or for work, and have no hope of returning to their hometowns (*Nishinihon Shimbun* 2012; *Asahi Shimbun* 2012). Family ties have also suffered. In many cases the father has had to remain in a contaminated area to continue working, with the rest of the family moving on to another prefecture. These developments have all had negative consequences for the human security of affected people.

The initial constraints, in the form of a lack of telecommunications facilities, electricity, gas and water supply, were compounded by a lack of food, roads being blocked by mud and debris, and children being forbidden to play outside because the ground was thought to be radioactive. Even one year after the top layer of soil was removed from playgrounds, children who played outdoors had to wear masks and could only be outside for a short time. Although the damaged reactors at the nuclear plant were shut down, problems associated with radioactive contamination remain.

Then there is the psychological fallout. The survivors were not prepared for the sudden onset of uncertainties: in half an hour they lost their families, friends, houses, companies, jobs, fields and fishing grounds. Many of the survivors were at a loss and in total despair after the disaster: they lamented that it was useless to go on living, that they could not sleep at night and that they did not have any prospect of restarting their businesses (*Fukushima Minpo* 2014).

Even those who were still able to work in their fields, supply fresh produce and catch fish could not sell the fruit of their labour at home or abroad, due to persisting safety-related rumours, despite there being a system to measure radioactivity and show prospective consumers whether produce is safe. More than three years on, some retail outlets organized special fairs to sell goods from Tōhoku, and some fish have been declared safe and are back in local shops as well as other shops nationwide. Nevertheless, people continue to come up against disaster-related constraints. Survivors, with a sense of guilt when remembering the deceased, are finding it difficult to have hopes and dreams for the future. Those who have been relocated elsewhere have a strong desire to return to their hometowns, but cannot. The occasional return visit is the best they are permitted. Not a few have opted for suicide, a choice reflecting their loneliness (*Fukushima Minpo* 2014). The psychological aspects of the survivors' lives urgently require attention, and that can be provided in a number of ways, including by involving them in cultural activities.

Coping with the triple disaster

In the interests of maintaining human security, societies invariably do their best to cope with, and reduce the impact of, conflicts and disasters by focusing on prevention, response and recovery. First, although it may be difficult to prevent natural disasters, certain measures can be taken to preserve the natural environment and so, by mitigating the effects of climate change, the human factor in natural disasters can be reduced. Second, in future, when preparing for a possible natural disaster, it should not be forgotten that educational drills do help, and can be the difference between life and death. In fact, it was found that those schools in Tōhoku which had held disaster drills incurred lower casualties, while those which had not had drills reported the loss of more students. One sad example concerns Okawa Primary School. In the tsunami it lost 74 (four still missing) of its 108 students and 10 (one still missing) of its 13 teachers (*The Star* 2012). The deaths were a result of the choice being made not to evacuate the school immediately. Moreover, there was no manual telling teachers where to take pupils in the event of a tsunami; otherwise, all the students and teachers might still be alive (Kikuchi 2011). As is noted in the Hyogo Framework for Action, it is essential to nurture 'a culture of disaster resilience' (UNISDR 2005, p. 7). Third, cultural activities have a large role to play in reducing adverse impacts on post-disaster recovery, as will be demonstrated below.

A people-centred approach

In the wake of the Great East Japan Earthquake, Japan faced two broad challenges. One was restoring the affected areas in such a way as to reduce the impact on lives and infrastructure, should another similar disaster occur. This will take time. The other challenge is reviving the lives of local populations, a task that requires urgent action.

In April 2011, the Japanese government set up the Reconstruction Design Council, an advisory panel of intellectuals set up to suggest how the affected areas might be rebuilt. The panel recommended that the focus be on local communities and that the steps taken include: financial compensation for disaster victims, including those suffering the effects of radioactive contamination from the nuclear power plant; clearing away the mud and debris; rebuilding breakwaters, dikes and harbour facilities so that fishing could be resumed; moving coastal villages to higher ground, if villagers agree; rebuilding infrastructure and facilities, including schools, hospitals and roads; rethinking how to use the land along the coast; curbing the effects of radioactivity; providing medical and nursing care as necessary; and reviving the local farming and fishing industries to create jobs (Reconstruction Design Council 2011, p. 11). The council's report particularly emphasizes that efforts should be geared to linking people to communities and to nature. It stresses that the government should not be seen by the citizenry as taking a high-handed approach but rather as facilitating reconstruction that reflects the wishes of local citizens, on pain of finding itself unable to revive the local communities.

Also mentioned in the report is the need for better-designed towns and villages, with appropriate flood-prevention earthworks and secure evacuation routes in addition to the coastal defence structures (Reconstruction Design Council 2011, p. 21). Reconstruction must take into account the possibility of another, similar disaster occurring and, while there is no failsafe protection for coastal towns and villages, they certainly can be rebuilt in such a way as to minimize damage from future disasters. In addition, the report recommends thorough evacuation education and drills, noting that most of the students at schools which had practised drills and had evacuation manuals had survived.

During recovery and reconstruction, the focus has to be on people. Although the term human security was not used in the report in connection with either phase, in substance it has been a fundamental aspect of the approach which has been adopted. The Design Council's findings were in line with the 1994 HDR in terms of supporting the reconstruction of community security and taking a people-centred, bottom-up approach (UNDP 1994, pp. 17–32). To date, however, the term has been used only in foreign policy contexts by the Japanese government, and I believe that the massive triple disaster of 2011 clearly demonstrates that it has a place in domestic policy discourse. This is amply reflected in the Reconstruction Design Council's recommendations, although it remains to be seen how the government interprets them and whether it will implement them.

The importance of culture: national responses

How did Japan respond to the triple disaster? In this section, I look at actual examples of cultural activities in the recovery areas, and consider – from a national perspective – how effective they have been for local residents. In the following section, I look at the same aspects, but from a local perspective. The triple disaster left many survivors in the midst of total devastation, with truly major losses and in total despair. Parents, husbands, wives, children, fishing boats, town halls and offices were gone. One fisherman lamented that he had lost everything – except his bank loan.

Many disaster survivors experienced a sense of guilt for having survived, and for not having saved family members. 'I killed [so-and-so], because I did not hold their hands tightly enough', they would say, or '[so-and-so] died, but it should have been me'. Many of the survivors blame themselves for the loss of their loved ones, and say such things as, 'I cannot forgive myself for having survived' (Washida and Akasaka 2012, pp. 130–8).

Seiichi Washida, a philosopher and former president of Osaka University, observes that, in a post-disaster setting, the material reconstruction of daily life and resumption of economic activities have always come first in Japan, with culture regarded as something to take care of last. In March 2011, however, the situation was very different. Artists flocked to the affected areas right after the earthquake, volunteering and helping in whatever ways they could, even performing for survivors in shelters. They found that, through a variety of art forms, they could engage the lonely and desperate (Washida and Akasaka 2012, pp. 130–8).

Initially, right after the disaster, the mood in Japan was such that, out of respect, cultural activities were cancelled: there were no concerts, exhibitions, festivals or theatrical performances. But this mood proved short-lived. In this section, I give concrete examples of how artists rushed to Tōhoku to support survivors through cultural activities; actors, singers and musicians, some originally from affected areas, were among the first to go to the devastated areas.

To cheer up schoolchildren who were depressed at finding many classmates missing, famed actor Masatoshi Nakamura visited his former high school in the Miyagi Prefecture town of Onagawa, which had been hit hard by the tsunami. The students said that meeting a celebrity had got them up on their feet again and shaken them out of total despair. Nakamura, together with singer Junichi Inagaki and eight other singers soon formed an NGO so that they could repair the musical instruments of children in the area. They did not want any child to have to give up their dream of becoming a musician, and believed they could help by repairing damaged instruments (*Asahi Shimbun* 2011a).

In April 2011, one month after the disaster, singer Muneyuki Sato, also from the affected area and known for his repertoire depicting the region's scenery, visited shelters in Ishinomaki to sing for the survivors. As they heard familiar songs, audiences were moved to tears and found their tension eased as they were shaken out of their apathy (*Asahi Shimbun* 2011a).

Sho Asano, born in Sendai City, Fukushima Prefecture, and a Tsugaru *shamisen* player since the age of five, also took his musical talent back to Tōhoku. A well-travelled performer who has given *shamisen* concerts outside Japan, Asano also played in other parts of Japan in a bid to raise funds to help support disaster survivors. Audiences were deeply moved at the sound of one *shamisen* he played that had been made using debris from the tsunami (Asano 2012).

Mostly, artists said that they had a strong desire to help the survivors, and thought the best way was to do this through their art (*Asahi Shimbun*, 2011a). Many artists performed around Japan to raise donations for recovery and reconstruction efforts. Among the artists from abroad, some cancelled their Japan tours, while others – such as conductor Zubin Mehta and singer Placido Domingo – came nevertheless and even visited disaster areas, a gesture that was truly appreciated by their audiences (*Nihon Keizai Shimbun* 2011a; Things Inspire 2011).

But the desire to help did not stop there. Athletes also played a part in helping survivors. Sumo wrestlers displayed their wrestling skills and served hot food to evacuees at shelters; professional soccer players engaged in games of soccer with children at shelters and taught many how to play the game. In addition, children were taken to play with famous football players whom they had seen on TV. In November 2011, Calvin Ripken, a well-known retired American major league baseball player, and Sachio Kinugasa, a retired Japanese baseball player, together visited the tsunami-ravaged area, travelling from Kesennuma City in Miyagi Prefecture to Rikuzentakata City and Ofunato City in Iwate Prefecture. These seasoned baseball veterans found themselves at a loss for words when they saw what was left of whole towns that had been washed away, leaving only rubble and mud that people were slowly removing. The two held baseball clinics for the children, who cherished their time with the famous former players. Ripken and Kinugasa noted that the attraction and power of sport had enabled the children to smile again, despite everything (*Asahi Shimbun* 2011b). When children are doing something they love, there is no room to think about anything else. Both veterans had wanted to help in their role as athletes, recognizing the joy that results from children's love of sport, and the importance of building ties with children (*Asahi Shimbun* 2011b).

The cultural activities presented here, which were conducted by individuals living outside the affected areas, represent only the tip of the iceberg, because so many cultural projects have been organized. The survivors interpreted the activities as a measure of how much people cared about them, and how willing they were to help. Many people appreciated these activities, finding them helpful in coming to terms with their bitter experiences, good for soothing their trauma, fear and loneliness, and key to finding new hope for the future. Yet, as also happens in the context of peace-building, some people did not welcome the cultural events. Sitting still, backs turned to the performers and, when in shelters, with blankets covering their heads, these people said they wanted material and financial – not cultural – help.

Cultural activities can play a role in disaster recovery when combined with efforts in other areas, so a comprehensive, multi-faceted approach is necessary. For example, people in Tōhoku were required to walk long distances to get almost everything and, as time passed, they began to feel they had been forgotten, due to the very slow pace of recovery. For these people, cultural activities played a major role, showing them that people in the rest of the country were thinking about them. But, after the completion of the visible phase of assistance – the bringing in of food and the rebuilding of infrastructure – there remained the long-term aspects of recovery. For this, the area continues to need outside support. Just as with peace-building efforts, cultural activities are well-suited to providing an important and neglected part of the required assistance: the means to rebuild the hearts and minds of the people.

The importance of culture: local responses

While the above examples involve groups and individuals who currently live outside the disaster area, it should be noted that survivors also took their own cultural initiatives. These match closely with human security's concern for grassroots action. In this section, I mention three local festivals, one region-wide festival, the successful re-establishment of a culturally significant and economically viable pottery-making practice and a museum exhibition.

The coastal village of Kitaura, Onagawa City, in Miyagi Prefecture, lost 58 of its 62 houses, so its 300 survivors were forced to evacuate to Akita Prefecture. There they were put up in a number of shelters, from which they were scattered in temporary housing and lost their sense of community. Each of the 15 fishing communities along the Onagawa coast has its own, separate lion dance, with distinctive choreography, rhythm and melody. Despite their despair, the villagers of Kitaura decided that in November 2011 they would hold their annual festival, featuring a lion dance. However, most of Kitaura's lion dance costumes and equipment had been washed away by the tsunami, and they were unable to borrow costumes or props, because their traditional music and lion's head are not like those of other villages. Villagers thus had to retrieve what they could from the mud, and supplement this with new additions.

In planning the festival, villagers were divided over whether they should perform the lion dance at all, since it is performed to mark happy occasions. However, because it is also performed for the repose of the souls of deceased humans, animals and insects, the villagers in the end agreed to put on the dance. It did much to boost the spirits of the local people, even those who had opposed having the festival and, through the traditional dances, villagers rediscovered their identity, bonds and solidarity as a group. After the festival, when they had danced at shelters and in old communities, the villagers felt relieved for the first time since the earthquake; they could now forgive themselves for having survived. Reminded of their roots and place in history, the survivors of Kitaura decided to build a new village on higher ground. Their consensus made them eligible for a national government grant for the construction of a new village.[3]

The tsunami-hit Fukushima Prefecture city of Sōma additionally suffered the impact of radioactive contamination from the nearby damaged nuclear power plant. With most of its houses and all but one of the trees in its famous pine forest washed away, the city was desolate. But it has a traditional festival, Nomaoi, hosted by three shrines and involving an old-fashioned horse race that dates back more than 1,000 years. Since the festival grounds were radioactive due to the nuclear power plant accident, and the village horses either had been washed away or were in a state of shock, much doubt surrounded the festival. In the end, however, it was held. The villagers prayed for the dead and for the revival of the area, giving them the strength they needed to make a bid for reconstruction (Mikasa 2012).

The significant role of culture in post-disaster psychological rehabilitation can also be seen in an example drawn from Kesennuma City, Miyagi Prefecture. A representative of the Kesennuma Kite Association, Masakatsu Kato, explained that the best kite-flying season is in winter, in February when the wind blows from Mt Anba to Kesennuma Bay. Nevertheless, he told the local people, he wanted to organize a kite festival after the earthquake. Many told him that it was wrong to enjoy kite flying after such a disaster and, moreover, the wharf used for the annual kite festival had been destroyed. Yet Kato went ahead. In May he organized a kite-flying event in the playground of a local elementary school. He received hundreds of kites from all over Japan for the children; youngsters in shelters and those who had been evacuated from their homes all came to fly kites. Kato saw the smiling eyes of the children and noted that the parents were happy to see their children playing again (*Nihon Keizai Shimbun* 2011b).

In addition to the revival of local festivals, six mayors in the devastated area organized a grand festival for July 2011, to wish the region a speedy recovery. Called Rokkonsai, the festival brought together representatives of six Tōhoku festivals in Sendai City, Fukushima Prefecture. The festival attracted a larger crowd than the organizers had expected, so a second festival was held in Morioka City, Iwate Prefecture, in June 2012. Attended by 243,000 spectators, it was the largest crowd ever to gather there. The streets were filled with a strong sense of solidarity for reconstruction, while some participants hoped that the festival would help revive tourism (*Nihon Keizai Shimbun* 2012b).

Local initiatives were not limited to festivals. The Fukushima Prefecture town of Namie is known for its Obori Sōma-yaki pottery that dates back to the late seventeenth century, when it was brought to Japan from the Korean Peninsula. The famed pottery is decorated with depictions of horses (representing the Sōma-han clan) and its tea cups feature a double-wall structure (to keep the contents hot, yet the cup cool enough to hold) and a blue crackle glaze. At the end of the Edo period (1615–1868), there were 100 kilns making Obori Sōma-yaki, of which Namie was the largest producer. Prior to the 2011 disaster, the town's annual exports to the United States alone totalled around ¥300 million. But, in the wake of the Fukushima Dai-ichi nuclear power plant accident, the town was declared an exclusion zone and pottery-making ceased. However, because potters did not want to extinguish their kiln fires – and, thus, end their pottery-making tradition – 22 of the potters decided to build a new kiln in Nihonmatsu City. Since they could no longer apply the natural glaze they had previously used, they spent the next year developing an artificial glaze that produces the same effect (*Nihon Keizai Shimbun* 2012a).

When Tsutomu Fujii, an artist in Morioka City, Iwate Prefecture, learned that his 1990 painting titled *Wind and Tide* that had been on exhibition in Rikuzentakata City, Iwate Prefecture, had been washed away by the tsunami, he decided to paint it again. Fujii agreed to exhibit it in a special exhibition, *Life and Land*, at the Ishigamino-oka museum in September 2011. It was not until a charter bus took 40 Rikuzentakata residents to see the painting that, observing the people's reactions, Saito, the director of the museum, realised the great power of art. He had learned that, after 9/11, almost all New York museums were closed – but the Metropolitan Museum of Art had remained open, a decision that had been strongly supported by New Yorkers. Saito recalls that the number of visitors to the Iwate museum in 2011 was more than for the previous year. He believes that, while art does not feed one, it is important for the hearts and minds of survivors (Akasaka 2012, pp. 338–341).

Masaru Kumagai, curator of the Rikuzentakata City museum, argues that cultural heritage items should be included in post-disaster reconstruction, and is working to repair relevant items that were damaged in the disaster. Unless local populations know the value of local cultural treasures, Kumagai argues, they will not have the drive to push for reconstruction. Although activities such as moving towns to higher ground, in preparation for any future tsunami, and building higher breakwaters should certainly be prioritized, physical reconstruction is not the end of the story. In Rikuzentakata, all but one of the trees in the pine grove along the beach were washed away. The surviving pine is now called 'the tree of hope', and has become a symbol of resilience. Kumagai makes the case that the pine forest should be replanted along the beach, as a windbreak and a tourist attraction. In the meantime, while the trees are gone, memories and stories about the pine trees must be kept alive at the museum (Akasaka 2012, pp. 42–6).

These are only a few examples of the local cultural initiatives taken. Although at the time some argued that it was inappropriate, immediately following the triple disaster, to conduct cultural events that would be enjoyed, others took the initiative to revive local traditional activities and events. These people addressed and overcame their sense of guilt for having survived, learned how to smile again, restored solidarity among villagers and generated new energy to work for the future. Although these activities must be combined with reconstruction, the aforementioned episodes suggest that the cultural side of recovery is an important part of the mental infrastructure on which a robust recovery is built.

Conclusion

The cultural events and projects discussed above, which were undertaken at national and local levels after the Great East Japan Earthquake, were of great value to the people in the disaster-affected areas of Tōhoku. Of course, during the post-disaster phases of response and recovery, addressing the material aspect of disaster management is essential. Nevertheless, in observing popular reactions to cultural activities, one can also see the impact such activities have on psychosocial aspects of the survivors' lives. Furthermore, there are clear similarities between the impact that cultural activities have in both post-disaster settings and peace-building efforts – the latter involving the following:

1 Helping local people to communicate with former foes in the interests of ultimately achieving reconciliation.
2 Turning despair into hope for the future.
3 Caring for and healing the psyches of survivors.
4 Rebuilding the self-respect and pride lost during conflict.
5 Empowering local populations.

It may seem that the first point – reconciliation – does not apply in the case of a natural disaster, insofar as the victims do not fight against enemies. However, reconciliation is often necessary when planning and rebuilding for the future in a post-disaster context, in addressing questions of responsibility and blame in connection with a disaster. The second point – turning despair into hope for the future – also applies in both peace-building and disaster settings. After all, populations hit by natural calamity and deprived of almost all their possessions will be in despair, and need to generate hope for the future if they are to recover from the damage and loss. Concerts, sporting events, and festivals enable them to find hope for the future.

Thirdly, like conflict survivors, disaster victims require adequate attention to be paid to their trauma, if their suffering is not to be long-lasting or lifelong. This is where cultural activities come into their own, as a medium to help heal psychological scars resulting from the loss of family and friends. Earthquakes and tsunamis make quick work of dissipating the self-respect and pride of victims – the fourth point. But in such cases, the restoration of items of local heritage, such as the pine trees of Sōma City, works wonders in restoring popular pride, as does the holding of local festivals. Cultural traditions invariably are couched in terms of feelings of pride and self-respect which, as I have recounted, led to a revival of solidarity among villagers in Tōhoku. Finally, empowering local populations, by means of sport or music-related projects, has been shown to encourage people, particularly children, to engage. Participating in cultural events has helped give people the strength they need to recover.

There is a structural similarity between peace-building and post-disaster reconstruction; in both cases, cultural activities have been shown to bring comfort and joy, as well as a moment in which to be moved and touched collectively. When members of an audience are moved by a cultural performance, they feel a sense of unity with the performers. When people talk and do things together, they feel that they can still take part in baseball games, soccer games or group singing. It is such experiences that allow people filled with despair to experience something new, or something they once knew. When survivors plan and hold festivals again, such as the theatre performances held in post-conflict settings, they find that they can enjoy these together, despite their unspeakable experiences, while in disaster situations they can regenerate the solidarity destroyed by the disasters. The power of culture lies in its ability to provide common ground for people emerging from conflict or disaster to share joy, which in turn empowers local people, who are the prime movers in any recovery.

Adopting a human security perspective allows us to address both the material and psychological dimensions of natural disasters. The people are paramount – in the affected areas and beyond – and the vulnerabilities revealed by the Great East Japan Earthquake indicate both that human security was endangered, and that a culturally-informed, people-oriented human security approach is required, if post-disaster reconstruction is to fully succeed.

Notes

1 Based on author interviews with UNESCO officials on 6 Sept. 2010.
2 Based on author meeting with Kuwayama on 23 April 2009, at Aoyama Gakuin University.
3 Story related to the author in February 2012 by Azusa Kamiyama, a student at Tōhoku University, who worked at Onagawa town hall.

References

Akasaka, N. (ed.) 2012, *Chinkon to saisei* (From the repose of the soul to its revival), Tokyo: Fujiwara Shōten.

Asahi Shimbun 2011a, 'Geinō pawā todoke' (Send the power of the arts), 23 April.

Asahi Shimbun 2011b, 'Fukutsu towa mae o mite susumu koto' (Resilience means to move forward), 18 Nov.

Asahi Shimbun 2012, 'Modoranu chōmin nayamu hibi' (Citizens not returning, filled with concern), Gifu edition, 28 Sept.

Asano, S. 2012, Remarks during a performance at Aoyama Gakuin University, Shibuya, Tokyo, 1 March.

Commission on Human Security 2003, *Human Security Now*, New York: CHS.

Collier, P., Elliot, V. L., Hegre, H., Hoeffler, A., Reynal-Querol, M. and Sambanis, N. 2003, *Breaking the Conflict Trap: Civil War and Development Policy*, New York: Oxford University Press.

Enarson, E. 2000, '"We will make meaning out of this": Women's cultural responses to the Red River Valley flood', *International Journal of Mass Emergencies and Disasters*, 18(1), pp. 39–42.

Fukushima, A. 2010, *Ningen no anzen hoshō* (Rethinking human security: responding to emerging global threats), Tokyo: Chikura Shōbo.

Fukushima, A. 2012, *Funsō to bunkagaikō* (Conflict and cultural diplomacy), Tokyo: Keio University Press.

Fukushima Minpo 2014, 'Shinsai kanrenshi: mienu shōrai hikan' (Disaster-related deaths: Pessimistic over future), 28 Sept.

Japan Reconstruction Agency 2014, *Japan Reconstruction Agency Report*, 28 March, http://www.reconstruction.jp, viewed Jan. 2014.

Kikuchi, M. 2011, 'Naze Okawa Shōgakkou dake ga daisanji ni nattanoka' (Why did only Okawa primary school have so many victims?), *Chūō kōron*, Aug.

Mikasa, A. 2012, 'Jisedai ni tsunagaru Sōma no samurai damashii' (Sōma's samurai spirit handed down to future generations), *Waraku*, July, pp. 18–23.

Ministry of Foreign Affairs 1999, 'Medium-term policy on official development assistance', www.mofa.go.jp/policy/oda/mid-term/1999/approach.html, viewed June 2012.

National Police Agency 2014, 'Damage Situation and Police Countermeasures associated with 2011 Tohoku district – off the Pacific Ocean Earthquake', 24 Oct., http://www.npa.go.jp/archive/keibi/biki/higaijokyo_e.pdf, viewed Jan. 2014.

Nihon Keizai Shimbun 2011a, 'Geijutsu no chikara shinjiru, Zubin Mehta' (Zubin Mehta believes in the power of art), 20 March.

Nihon Keizai Shimbun 2011b, 'Kesennuma no sora ni agaru tako' (Kite flying in the sky over Kesennuma city), 3 May.

Nihon Keizai Shimbun 2012a, 'Dentō no toki fukkatsu-e' (Reviving traditional pottery), evening edition, 10 May.

Nihon Keizai Shimbun 2012b, 'Tōhoku rokkon sai no chosen wa tsuzuku' (The challenge of the Rokkon festival continues), evening edn, 19 June.

Nishinihon Shimbun 2012, '"Gonen wa modorenai" to Namie-chō: Fukushima genpatsu jikōgo zen-chō hinan' ('We cannot return for at least the next five years,' says the Namie town mayor: All citizens had to evacuate after the Fukushima Nuclear Power Station accident), 10 Sept.

Reconstruction Design Council 2011, 'Towards reconstruction: Hope beyond the disaster', Report to the prime minister in response to the Great East Japan Earthquake, 25 June, http://www.cas.go.jp/jp/fukkou/english/index.html, viewed Jan. 2012.

The Star 2012, 'Japan tsunami's wake leaves grieving mothers still searching for children in school disaster zone', 17 Feb., www.thestar.com/news/world/article/1133603, viewed Oct. 2012.

Things Inspire 2011, 'Placido Domingo sings to comfort Japanese', web blog post, 11 April, http://thingsinspire.blogspot.com/2011/05/placido-domingo-sings-to-comfort.html, viewed Oct. 2012.

UNDP 1994, *Human Development Report*, New York: Oxford University Press.

UNGA 2005, *2005 UN World Summit Outcome*, A/RES/60/1, 24 Oct., http://mdgs.un.org/unsd/mdg/Resources/Attach/Indicators/ares60_1_2005summit_eng.pdf, viewed Aug. 2013.

UNGA 2012, 'Follow-up to paragraph 143 on human security of the 2005 World Summit Outcome', 25 Oct., http://www.un.org/ga/search/view_doc.asp?symbol=%20A/RES/66/290, viewed Aug. 2013.

UNISDR 2005, *Hyogo Framework for Action 2005–2015: Building the Resilience of Nations and Communities to Disasters*, A/CONF.206/6, 22 Jan., http://www.unisdr.org/2005/wcdr/intergover/official-doc/L-docs/Hyogo-framework -for-action-english.pdf, viewed Sept. 2012.

UNSG 2010, *Human Security: Report of the Secretary-General*, A/64/701, 8 March, http://www.un.org/ga/search/view_doc.asp?symbol=A/64/701&Lang=E, viewed Sept. 2012.

UNSG 2012, *Follow-up to General Assembly Resolution 64/291 on Human Security, Report of the Secretary-General*, A/66/763, 5 April, http://www.un.org/ga/search/view_doc.asp?symbol=A/66/763&referer=http://search?je=utf8&Lang=E) , viewed Sept. 2012.

Washida, K. and Akasaka, N. 2012, *Tōhoku no Shinsai to sōzōryoku : wareware wa nani o owasareta no ka*, (Imagining the Tōhoku disaster: the issue of blame), Tokyo: Kōdansha.

10 What role for nuclear power in Japan after Fukushima?

A human security perspective

Paul Bacon and Mai Sato

This chapter brings a human security lens to bear on the energy mix question in post-Fukushima Japan. In particular, two of the four elements of human security identified in the 1994 *Human Development Report* (*HDR*), prevention and people-centeredness, are mobilized. We trace developments in Japan's post-Fukushima nuclear politics through the demise of DPJ rule to the advent of the LDP government, and evaluate the current nuclear energy strategy of the Abe administration. Using a human security framework, we consider the economic security dimension of the arguments for and against the use of nuclear power, and weigh the result of this consideration against a concern with the six other elements of human security identified in the 1994 *HDR*. We conclude that the risks and threats to human security engendered by the use of nuclear energy outweigh any benefits that could reasonably be argued to accrue from its use. The notion of prevention, so central to the concept of human security, performs a further 'trumping' function, in leading us to put a premium on the downside risk of the use of nuclear energy.

We also argue that the Abe administration's energy strategy is not sufficiently people-centered, in that it goes against the wishes of the Japanese public. Whilst it is true that Abe and the LDP did run successfully on a pro-nuclear platform in the elections of December 2012 and July 2013, this does not necessarily mean that he received a mandate for his nuclear energy strategy. He focused his campaign almost exclusively on economics, and the election can also be characterized as a punishment of the DPJ rather than a full endorsement of the LDP. Furthermore, in the three years since the Fukushima crisis first manifested, the Japanese public has consistently shown that it is opposed to nuclear energy in a series of opinion polls, and through protest and demonstrations. Interestingly, the Japanese public also expressed opposition to nuclear power in a deliberative poll commissioned by the DPJ. The weight of evidence, the general election result notwithstanding, suggests that there has been substantial and sustained opposition to nuclear power in Japan. A genuinely people-centered approach should reflect the resilience of this core opposition, and address the issue of how to empower Japanese public opposition to the nuclear option.

We conclude by noting that, at the time of writing, the Abe administration is pushing for nuclear reactor restarts without adopting a clear position in the larger debate on the energy mix question in Japan. This is a tactically sound short-term strategy, which will lead to the restarting of some nuclear reactors in 2014. But adopting this strategy also means that the longer term debate about the ideal energy mix will remain in play at the next elections in 2016. This means that there is still time to mobilize public and political opposition to a strong role for nuclear energy in the medium to long term.

Human security and the emphasis on prevention

The 1994 *HDR* identifies four elements of human security and seven dimensions. The four elements are: that human security is a universal concern; that there should be an emphasis on preventing the emergence of threats to human security; that human security should be people-centered; and finally that human security threats are interrelated (HDR 1994, pp. 22–3). In this chapter we focus on the last three of these in some detail. Human insecurity is most visible after a disaster has occurred, but it is best mitigated through preparation beforehand. But what would a commitment to prevention of human insecurity in the context of natural disaster entail in a Japanese context, and how does this relate to Japanese energy policy? Many (Kingston 2013a; Morton 2012; Noyes 2011) have suggested that maintaining a substantial nuclear power-generating capacity in the future increases the likelihood of sudden and hurtful disruption, and exposes people to critical and pervasive threats. Others have argued that nuclear energy is the only economically viable option, and that the risks associated with its use have been exaggerated. To address this question of prevention of future human insecurity, it is necessary to explore the recent historical context, look in more detail at the debates surrounding the future Japanese energy mix and evaluate the threats to human security on both sides of the ledger.

Background and historical context

Early in 2011, nuclear energy accounted for almost 30 per cent of Japan's total electricity production. There were plans to increase this to 41 per cent by 2017, and 50 per cent by 2030. By mid-May 2011, as a result of the March triple disaster, only 17 out of Japan's 50 remaining nuclear power reactors were in operation. After May 2011 the number of operating reactors steadily decreased to zero. In the summer of 2011 tough energy conservation measures were applied, leading to a 12 per cent reduction in power consumption (relative to 2010) in August, and more significantly, a reduction in peak demand of 18 per cent, exceeding the government target of 15 per cent (World Nuclear Association 2013).

In March 2012 Japan's Atomic Industrial Forum (JAIF) said that 35 nuclear reactors might be subject to extended outage, and that the fate of some others was in doubt. In mid-April 2012, after a series of high-level meetings, the Japanese government approved the restart of Kansai Electric's Oi 3 and 4 reactors, and urged the Fukui governor and the Oi mayor to endorse this decision. They restarted in July 2012. Without these units, it was predicted at the time that significant (12 per cent) electricity shortages were likely in summer peak periods (World Nuclear Association 2013).

In July 2011 an Energy and Environment Council (Enecan) was set up by the Cabinet Office as part of the National Policy Unit to make recommendations on Japan's energy future to 2050. Its initial review was to recommend that nuclear power's contribution to electricity be targeted at 0 per cent, 15 per cent or 20–5 per cent for the medium term – a 36 per cent option was dropped. Enecan's 'Innovative Energy and Environment Strategy' was released in September 2012, recommending a phase-out of nuclear power by 2040. In the short term, reactors currently operable but shut down would be allowed to restart once they gain permission from the incoming Nuclear Regulatory Authority, but a 40-year operating limit would be imposed. Reprocessing of used fuel would continue. Enecan promised a 'green energy policy framework' focused on burning imported gas (LNG) and coal, along with expanded use of intermittent renewables. However, at the end of 2012 the new LDP government promptly abolished Enecan, along with the National Policy Institute, so that METI's Advisory Committee for National Resources and Energy became responsible for formulating energy plans (World Nuclear Association 2013). In October 2012, the new Nuclear Regulatory Authority (NRA) which had taken over from the Nuclear and Industrial Safety Agency and the Nuclear Safety Commission, announced that nuclear power plant restart reviews would comprise both a safety assessment by NRA and the briefing of affected local governments by the operators. The assessments would be based on safety guidelines formulated by the NRA in July 2013 after public consultation.

The case in favor of nuclear power

In December 2012 the LDP, campaigning on a pro-nuclear platform, won a decisive victory in national elections for the Diet's Lower House, winning 294 out of 480 seats, and indicated that it would restart idled nuclear power plants. The new government said it would take responsibility for allowing reactor restarts after the NRA issued new safety standards and confirmed the safety of individual units. In the July 2013 elections for the Diet's Upper House, the LDP won 115 seats out of 242, which consolidated the LDP position and role in reviving the economy, including restoring power supplies. The DPJ, which had adopted a policy of abandoning nuclear power by 2040, won only 59 seats. The LDP won a seat in every constituency with a nuclear power plant. In Fukushima Prefecture the LDP candidate polled more than twice as many votes as the DPJ candidate. In Fukui Prefecture, where the Kansai Electric Power Company has 11 units, the Japan Atomic Power Company has two units, and the government has the Monju prototype breeder reactor, an LDP candidate beat the DPJ contender, 237,000 votes to 56,000. From a pro-nuclear perspective, this amounts to a clear popular vindication of the Abe administration's energy policy (World Nuclear Association 2013).

There are strong economic motives for moving towards reactor restarts. The JAIF has claimed that increased fuel imports have been costing about

¥4 trillion ($40 billion) per year (METI puts total fossil fuel imports at ¥9 trillion in the financial year 2013). The trade deficit in 2012 was ¥6.9 trillion ($70 billion), and the same is expected in 2013. The total trade *deficit* from March 2011 to end of the 2013 financial year in March 2014 is estimated at ¥16.4 trillion ($168 billion), compared with previous surpluses of at least ¥2.5 trillion per year. Power generation costs increased 56 per cent in the financial year 2012. Losses across the utilities are about ¥1 trillion per year. The Ministry of Economy, Trade and Industry (METI) said in April 2013 that Japanese power companies had spent an *additional* ¥9.2 trillion ($93 billion), up to that point, on imported fossil fuels since the Fukushima accident.

The debate between nuclear and fossil fuels raises interesting issues with regard to the issue of environmental security. According to the pro-nuclear World Nuclear Association, carbon dioxide intensity from Japan's electricity industry rose again in financial year 2012, reaching levels 39 per cent greater than when the country's nuclear reactors were operating normally, and taking the sector far beyond established climate targets. About 100 million tonnes per year more CO_2 is being emitted than when the reactors were operating, adding 8 per cent to the country's emissions (World Nuclear Association 2013). One of Japan's climate change goals had been for the electricity sector to reduce carbon intensity by 20 per cent from 1990 levels. These changes resulted in Japan revising its goals and making more modest commitments to reducing domestic greenhouse gas emissions at the 19th Conference of the Parties to the UN Framework Convention on Climate Change (COP19) held in Warsaw in November 2013. Japan was criticized for this revision by other developed and developing countries (*Yomiuri Shimbun* 2013).

One way to mitigate the grim trade-off between warmer or more irradiated seas is of course to grow renewables as a proportion of power generation. The green energy industry was greatly boosted by the distrust and fear of nuclear power triggered by the 2011 earthquake and tsunami disaster that crippled the Fukushima Dai-ichi nuclear power plant. The then-ruling DPJ introduced a renewable energy feed-in tariff system in July 2012. This system obligates utilities to purchase electricity generated by solar and wind plants at predetermined prices. The system was an attempt to boost Japan's renewable energy production, which accounted for less than 2 per cent of total power generation at the time, to 30 per cent, to replace the share that nuclear power once claimed. *The Economist* (2013) states that renewables now represent 10 per cent of the energy mix.

While renewables have grown since 3/11, there is also some evidence that Japan's renewable energy industry has recently lost momentum. According to a February 2013 nationwide survey by the Japan Renewable Energy Foundation, 34 of the 79 solar energy producers who responded said they had given up on at least one solar power project. Roughly 45 per cent of those respondents cited difficulties in land procurement, followed by 25 per cent who said they had problems joining the power grid. A brief discussion of the situation in Hokkaido illustrates this point well. Hokkaido is a key battleground in the broader struggle to increase the proportion of power generated by renewables, with solar power potential in the south of the island and wind power potential in the north. According to Toru

Suzuki, chairman of the non-profit Hokkaido Green Fund 'no growth target for renewable energy would be feasible without Hokkaido' (Fujisaki 2013). It is therefore highly significant that in April 2013, the Hokkaido Electric Power Company said it would only purchase a total of 400 megawatts of electricity as part of the feed-in tariff system from the so-called mega-solar power plants, each with a generation capacity of 2 megawatts or more. That amounts to turning down as many as 70 per cent of the 87 applications to sell it solar power, filed up until March, with a combined output capacity of 1.568 gigawatts (Fujisaki 2013).

A Hokkaido Electric official attempted to justify this decision by claiming that their power grid has limited capacity, and that accepting too much power from solar plants, where output levels fluctuate wildly depending on the weather, would compromise the stable supply of electricity. Whilst there is an element of truth to this claim, the problem is that the regional utility's decision to limit its purchases of solar power cannot be attributed to grid capacity alone. The decision was taken in large part due to Hokkaido Electric's hopes and expectations that all three idled reactors at its Tomari nuclear power plant will eventually go back online (Fujisaki 2013). Power demand in Hokkaido at its minimum is just 3 gigawatts. The three Tomari reactors have a combined power generation capacity of 2.07 gigawatts, leaving a difference of roughly 1 gigawatt if they are returned to service. But if utilities revert to relying on nuclear power to levels before the Fukushima disaster, that could leave very little room for the emerging renewable energy industries to grow. Or, to put it another way, Hokkaido Electric has decided to limit the take-up of renewable energy to almost 25 per cent of its existing capacity in the prefecture, by imposing a 400-megawatt cap on 1.568 gigawatts of potential.

The community of electricity utilities, bureaucrats, academics and heavy industry in Japan known as the 'nuclear village' has predictably been urging the LDP to restart the reactors. But there are other forces pushing Japan in a pro-nuclear direction. These include the US, and also Japan's desire to make money from the nuclear industry worldwide. Prime Minister Abe, former Governor of Tokyo Shintaro Ishihara and LDP General-Secretary Shigeru Ishiba have all identified the benefits of retaining nuclear energy as a way of keeping Japan's nuclear weapons options open. *The Tokyo Shimbun*, among others, has claimed that the US government demanded that no Cabinet decision be made endorsing a phasing-out zero option (*Tokyo Shimbun* 2012). This is partly because it is also in the US geostrategic interest that Japan is seen to retain the nuclear weapons option, but another US motive is that Japan having large amounts of plutonium without a clear and credible plutonium-use plan sets a bad example to other countries, and raises questions about the consistency of the non-proliferation regime (Kingston 2013a). The US also benefits greatly from Japanese nuclear technology, and would therefore lose out if Japan were to abandon nuclear energy.

As well as fiercely defending nuclear power generation at home, Abe has emerged as a champion of the export of Japan's nuclear technologies and expertise. Abe's government helped secure a contract to build nuclear reactors in Turkey, and Abe himself has recently visited Saudi Arabia, India and Central Europe to promote Japanese nuclear capabilities. In late March, a group representing the

JAIF also visited the Sizewell nuclear power plant in the UK. The JAIF's members include power utilities and manufacturers dealing in nuclear technologies. There are plans to build two more nuclear reactors on the grounds of the Sizewell site. Japanese companies hope to benefit from the Abe administration's overseas sales pitch. Hitachi, for example, has explicitly endorsed Abenomics, and said that it hopes to boost the annual sales of its nuclear business division from the current ¥160 billion ($1.64 billion) to ¥360 billion by 2020 (Fujisaki 2013).

Based on the new safety guidelines which were introduced in July 2013, four of Japan's ten utilities have applied to restart twelve reactors, and the decision about restarting these reactors will happen in early 2014. If the NRA approves all of these applications for restarts, they could provide 10 per cent of Japan's electricity-generating capacity, and this would likely create additional momentum for further restarts (Kingston 2013a). It is of course a major priority for the Abe administration to get as many restarts as it can approved as soon as possible, and it has been putting significant pressure on the NRA. Kingston argues that, if the government does get some restarts, it is 'hard to imagine that the compelling logic of the green revolution will trump the hard-nosed institutional interests and power of the nuclear village' (Kingston 2013a). Although Abe ran for election on a pro-nuclear ticket, in contrast to the DPJ he has cleverly avoided being drawn into the larger debate about the ideal energy mix, and has not committed himself to clear nuclear targets. He has instead suggested that it may take ten years (*The Economist* 2013; Squassoni and Gorman 2013) to develop an appropriate energy policy, and has thereby successfully 'kicked the can down the road' (Kingston 2013a). In the meantime, from his pro-nuclear perspective, he has three years to press for the restart of as many reactors as possible. Abe is hoping to be able to ride out the current high levels of opposition to nuclear energy, and generate momentum, as unobtrusively as possible, for the re-establishment of nuclear energy's former and familiar position of strength. If he is able to ride out this storm, Abe will effectively be 'locking-in' nuclear energy for the foreseeable future.

Much depends on how many restarts Abe is able to get. Many anti-nuclear commentators were hoping that there would be none, and that any restarts would constitute a defeat of sorts. However, this was always a somewhat unrealistic perspective. There will be some restarts, but it is key to limit these, and brand them as a short-term exercise to manage emergency shortfalls in energy supply. At this writing, we can only speculate as to how many restarts there will be. The pro-nuclear World Nuclear Association (2013) estimates that restarts could proceed at a rate of 10 per year, boosted by a mooted increase in the number of NRA assessors. The CSIS estimates that the maximum pace would be six assessments per year which, even if all were positive, would require eight years for all restarts (Squassoni and Gorman 2013). Others have suggested a much lower figure; for example, Kazuhiro Ueta, a renewable energy specialist with the Japan Renewable Energy Foundation, has suggested that as few as 12–15 reactors could be restarted (*The Economist* 2013).

The case against nuclear power

Having discussed the forces pushing towards retaining nuclear energy in Japan, we would now like to consider the case against. As we will argue in more detail below, there is significant opposition to the nuclear option within the Japanese media and civil society. *The Japan Times* (2013) made its position clear in a strongly worded editorial calling on the government to 'work out a policy as soon as possible that shows a timeline for ending Japan's reliance on nuclear power generation'. One of the main objections to the continued use of nuclear energy is that there is currently no established technology to safely store high-level waste from nuclear plants for long periods of time. It can take tens of thousands of years for the radiation emitted by nuclear waste to drop to safe levels, and it is environmentally dangerous to store such waste on a semi-permanent basis and leave the problem to future generations to solve (*Japan Times* 2013).

The experience of the extremely hot summer of 2013 shows that all but one power company in Japan were able to meet electricity demands without relying on nuclear power. This disproves the predictions of the World Nuclear Association, cited earlier, for example. Only one company, KEPCO, used nuclear power during this period, and KEPCO and Kyushu Electric Power Company were the only companies to experience demand for electricity that reached or surpassed 95 per cent of supply capacity, and even then only for a few days (*Japan Times* 2013). *Japan Times* also argued that the government should cease its present policy of building large-scale nuclear and thermal power generation plants in a limited number of places, and should instead build many small-scale power generation facilities that use renewable energy sources near towns and cities across Japan. Emphasis should also change from increasing power generation capacity to reducing demand for electricity. Promising examples that this is possible include the fact that the general public was able to exceed the government target for reduced consumption in the summer of 2011. Tokyo has been able to maintain this reduced consumption for longer, having slashed it by a tenth since 2011, according to the Japan Renewable Energy Foundation (*The Economist* 2013).

Nuclear reactor restart applications are the highest profile issue in Japan's energy sector, but other important issues include proposals for structural changes to the power sector, and post-3/11 renewable energy incentives. Deregulation poses a threat to nuclear power development, since competition among service providers encourages shifts toward supply sources with cheaper startup costs. Utilities fear that the reduced electricity rates which are predicted under deregulation would prevent them from recovering the stranded costs associated with high startup investments in nuclear power (Aldrich et al. 2013). In the post-Fukushima era the Japanese government has renewed discussion of deregulation, as a means of enabling renewable energy entry into the power market, and has distributed generation to foster local energy supply stability.

However, in a positive development for those in the anti-nuclear lobby, there is legislation in the pipeline which is intended to liberalize the energy market. First, a new centralized, independent operator, the Organization for Nationwide

Coordination of Transmission Operators, will coordinate supply and demand and plan regional interconnection power lines. Second, the wholesale and retail markets will be fully liberalized to allow for free competition among suppliers, establishing wholesale trading markets for electricity. Third, generation, transmission and distribution will be legally unbundled, and retail electricity rates will be fully liberalized. If the policy recommendation is followed, these three stages will be implemented in 2015, 2016 and between 2018 and 2020, respectively (Aldrich et al. 2013). While these reforms could enable new entrants to supply power to the grid, namely renewable energy producers, questions have arisen regarding the impact of reforms spread over a relatively long time frame. According to Aldrich, several reporters for major media outlets have mentioned a rumor that the private utilities agreed to various deregulatory measures in exchange for central government support for nuclear restarts (Aldrich et al. 2013). If true, this dimension of Abe's energy policy represents something of a hedge; buttressing the position of the big utilities, but allowing for the possibility of greater competition.

As the Abe Cabinet proceeds with promised Abenomics reforms, balancing economic recovery with perceived energy security concerns is a priority. Discussion of regulatory reforms in the energy sector will proceed amidst continued investment in renewables. The government will face continued pressure from large industrial users to restart nuclear reactors, but anti-nuclear voices in the public and media will continue calls to reduce reliance on nuclear power. The Abe Cabinet has officially postponed big decisions on energy policy for several years, but short-term decisions on regulatory reform, renewables promotion and reactor restarts will shape the crafting of a new basic energy plan and Japan's energy future (Aldrich et al. 2013).

Evaluation: the priority of prevention

For anti-nuclear commentators, the triple disaster is one of the most important events in Japan's recent history, an event that should be seen as a fundamental game-changer: Kiyoshi Kurokawa, an academic who headed a commission investigating the disaster on behalf of the Japanese parliament, thinks that Fukushima has opened the way to a new scepticism about an ageing, dysfunctional status quo which could bring about a 'third opening' of Japan comparable to the Meiji restoration and the American occupation after 1945 (*The Economist* 2013). As can be seen from the previous discussion, it is not clear whether Kurokawa's hopes will be realized.

Having discussed some details of the situation within Japan, it is also valuable to look at some of the more general arguments that can be made against nuclear power, Oliver Morton, writing in *The Economist* in a 2012 Special Report on Nuclear Energy, claimed that:

> [i]n *liberalised* energy markets, building nuclear power plants is no longer a commercially feasible option: they are simply too expensive. *Existing* reactors can be run very profitably; their capacity can be upgraded and their lives

> extended. But forecast reductions in the capital costs of new reactors in America and Europe have failed to materialise and construction periods have lengthened. Nobody will now build one without some form of subsidy to finance it or a promise of a favourable deal for selling the electricity.
>
> (Morton 2012, my emphasis)

In 2010 the world's installed renewable electricity capacity outstripped its nuclear capacity for the first time. That does not mean that the world got as much energy from renewables as from nuclear; reactors run at up to 93 per cent of their stated capacity whereas wind and solar tend to be closer to 20 per cent. However, it is also worth noting in Japan spending on solar projects has doubled to $20 billion, for a market share of some 10 per cent, equivalent to the energy capacity of six nuclear reactors. Renewables are intermittent and take up a lot of space. *Unsubsidized* renewables can currently displace fossil fuels only in special circumstances. But nuclear energy, which has received large subsidies in the past, has not displaced much in the way of fossil fuels either. And nuclear energy is getting more expensive whereas renewables are getting cheaper (Morton 2012)

Thomas Noyes, writing in the *Guardian*, has suggested that, in the case of wind and solar power, it is possible to see the cost curve dropping to the break-even point in the near future. Nuclear power, however, is not as cost-effective as it is often claimed to be, and is only becoming more expensive after Fukushima. This is due to the daunting financial prospect of cost overruns, waste disposal and extended shutdowns. Large-scale disasters like Fukushima, however rare, are colossally expensive, as well as dangerous (Noyes 2011). Potential investors refuse to assume that risk, and if cold, calculating capitalist logic does not actually bring us to a pro-nuclear conclusion, and moral considerations certainly do not, then why should political considerations dictate that we choose the nuclear option? It should be noted here, in the interests of balance, that all forms of energy production require substantial government subsidy. But for opponents of nuclear energy, this is also precisely the point; proponents of nuclear energy base their arguments on the supposedly clear economic superiority of the nuclear option, when in fact these claims do not hold water on more detailed inspection.

As we have seen, it has been expensive for Japan to import fossil fuels in the past three years. But on which side of the ledger should these costs go? The nuclear industry likes to recognize this extra spending as part of its case that nuclear energy is cheaper. However, it could be argued that these amounts have had to be paid precisely because of the risks of nuclear, and the fact that it is necessary to import extra fossil fuels on unfavorable terms as a result of reactor failures and health risks. As *The Economist* (2013) notes, these costs have been exacerbated by a weaker yen and higher oil prices. One could strengthen this argument still further by noting that Japan's pre-Fukushima reliance on expanding nuclear power meant that, after the accident happened, Japan did not have a very diverse energy portfolio – it had neglected to develop renewables. If Japan had started working on sustainables 10–20 years ago, the amount of fossil fuel it would need to import now would be more limited.

When we do turn to other, non-economic dimensions of human security, we can also see that there is the potential for significant damage to our environmental security, and to our health security. In the event of failure, reactors leak radiation into the sea, soil and atmosphere, and also, therefore, into the food chain. For many, these considerations would outweigh the economic arguments, even if it could be shown that there is a strong economic case for nuclear energy. In this sense, although one should not trivialize such an important issue, TEPCO's 'grim comedy of errors' (DeWit and Hobson 2013) in its handling of the Fukushima Dai-ichi plant is a public relations gift that keeps on giving for the anti-nuclear lobby. TEPCO's travails in this and other cases provide compelling evidence of the nuclear village's unwillingness or inability to manage nuclear safety responsibly. This inability to mitigate threats to health and environmental security demonstrates that inadequate preventative and reactive measures have been and are being taken, which in turn highlights the ongoing risks of nuclear energy.

It may be useful at this point to return to human security's emphasis on preventing and addressing insecurities as part of normal, everyday politics. As the 2003 Commission on Human Security (CHS) report explains, 'the aim of human security is … to build a protective infrastructure that shields all people's lives from critical and pervasive threats' (p. 85). Here it may be useful to remind ourselves that 'in the darkest moments of [the March 2011] nuclear accident, Japanese leaders did not know the actual extent of damage at the plant and secretly considered the possibility of evacuating Tokyo, even as they tried to play down the risks in public' (Fackler 2012). The risk engendered by rare events such as major earthquakes and tsunamis undermines the economic arguments in favor of investing in nuclear energy, as we have seen. But nuclear energy also carries precisely the kind of risks and threats to human security which the 1994 *HDR* appeals to us to avoid if we can. It is difficult to imagine many things more costly and disruptive than the evacuation of one of the largest cities on the planet. Such an evacuation would threaten the economic, personal, health and community security of those affected. If we view events through a human security lens, and take seriously a commitment to prevention, the logical conclusion would be to discount nuclear energy as carrying too much risk to the fundamental condition of human security outlined in the 1994 and 2003 human security reports cited above.

Human security requires a people-centered approach

Opinion polls and protests

In this section, we consider the people-centered element of human security in the context of post-Fukushima nuclear politics. People-centeredness entails that we think of human security threats from the perspective of the people, and that we also take into account the views of the people in arriving at conclusions and deciding on policy solutions. As the 2003 CHS report puts it:

> Empowerment is important because people develop their potential as individuals and as communities. Strengthening peoples' abilities to act on their own behalf is also instrumental to human security. People empowered can … mobilize for the security of others. Supporting people's ability to act on their own behalf means providing education and information so that they can scrutinize social arrangements and take collective action. It means building a public space that tolerates opposition, encourages local leadership and cultivates public discussion.
>
> (CHS 2003, p. 11)

What does this mean in a Japanese context? One way of considering this issue is looking at Japanese public opinion on the nuclear energy question. According to *Asahi Shimbun*, 94 per cent of the Japanese public believe that the Fukushima accident has not been brought under control, 31 per cent want to abandon nuclear energy as soon as possible and an additional 54 per cent support a gradual phase-out (Kingston 2013b). A poll conducted by Jiji Press in June 2013 showed that 58.3 per cent of Japanese do not support the export of Japanese nuclear technologies and expertise, more than double the 24 per cent who do support such exports. Even among LDP voters, opponents of nuclear exports exceed supporters (43.2 to 40.4), according to the same poll (Jiji 2013a). Furthermore, a majority of the Japanese public (55 per cent) remains opposed to restarting nuclear power plants (Sieg 2013). In July, according to a post-election poll by the *Mainichi Shimbun*, only 25 per cent of victorious LDP *candidates* said that they viewed nuclear plants as necessary to Japan's well-being (Squassoni and Gorman 2013). Former LDP Prime Minister Koizumi, a highly respected figure in Japanese politics, has also famously pitched into the nuclear debate. In November 2013, in a high-profile speech at Japan's national press club, Koizumi argued that 'zero nuclear is the best choice' and that such a policy should be implemented 'immediately' – meaning that plants that have been shut for safety evaluations should not be restarted, even as a stopgap measure while alternative energy sources are developed (Soble 2013). At the local level, fewer than half (49 per cent) of mayors say that they would approve reactor restarts, even if certain safety conditions are met (Squassoni and Gorman 2013). Each of these polls demonstrates substantial opposition amongst the Japanese public, prefectural mayors, LDP voters, and even within the higher reaches of the LDP itself.

There have also been a number of substantial demonstrations against nuclear power. On 16 July 2012, an estimated 75,000 people gathered near a Tokyo park in what was described by NHK, Japan's national broadcaster, as Tokyo's largest anti-nuclear event yet. Recent demonstrations signaled a fundamental change in attitudes, as relatively few Japanese had been willing to take political issues to the streets since the 1960s (Dickie 2012). An estimated 20,000 people also rallied against nuclear power on 13 October 2013, in the largest demonstration since Japan's last active reactor went offline for maintenance in September 2013 (Jiji 2013b). Participants in this event had also gathered more than 8 million signatures against the government's plan to restart nuclear power plants after the 2011 disaster at the Fukushima nuclear power plant (United Press International 2013).

The government-commissioned deliberative poll

As well as these opinion polls and public activism, when in power the DPJ government, in an unprecedented move, commissioned a deliberative poll on energy security. Although the various opinion polls and public protests have rightly been given a great deal of attention, we argue that the deliberative polling exercise should have received more attention than it has because it provides a robust guide to public opinion. This poll was a very instructive exercise, one that further challenges Abe's claim that his mandate includes restarting reactors and promoting nuclear energy. The main finding of the poll is that a majority of the participants were opposed to future reliance on nuclear energy at the beginning of the poll, and became even *more* committed to this original position after the provision of information and opportunities for debate about nuclear energy.

Deliberative polling is an innovative social survey technique, designed to identify the considered attitudes and policy preferences that people express when they have been given the time and information to consider an issue fully. Like conventional attitudinal research, deliberative polling relies on structured surveys with a random sample. But unlike conventional research, deliberative polling requires participants to be assembled in one place to learn about a particular topic, discuss and exchange opinions on the issue, and answer pre- and post-deliberation surveys. It is a mixed-method survey producing both quantitative and qualitative data that allow a deeper examination of opinion and attitude.

Fishkin (1997, p. 162) describes the principle and the utility of deliberative polling in the following manner:

> The deliberative poll is unlike any poll or survey ever conducted ... A deliberative poll is not meant to describe or predict public opinion. Rather it prescribes. It has a recommending force: these are the conclusions people would come to, were they better informed on the issues and had the opportunity and motivation to examine those issues seriously. It allows a microcosm of the country to make recommendations to us all after it has had the chance to think through the issues.

Deliberative polling stems conceptually from Habermas's theory of communicative action (Habermas 1984, 1987), which aims to elevate public opinion to 'public judgment' (Yankelovich 1991) and to elicit views of 'ideal citizens' (Fishkin 1997, p. 162) through deliberation.

Every deliberative poll differs slightly due to what is practically feasible on each occasion, but there are several essential processes that all deliberative polls entail. Potential participants are contacted normally using a random probability sampling technique to ensure representativeness of the general public. Out of those contacted, those who agreed to take part in the poll are screened using a questionnaire. After the completion of the survey, briefing documents are given about the topic of the deliberative poll. Participants are asked to go through the material in their own time before the deliberation day. Particular attention is paid to providing objective and balanced information in a simple manner that is brief and easy to understand. The actual event takes place later, when all participants are gathered for one or two days to discuss and exchange views, and listen to experts. This is done through small discussion groups overseen by moderators. Participants also take the same – or a very similar – survey at the beginning and at the end of the event to measure changes in attitudes.

In August 2012, 'a microcosm of the Japanese people' (CDD 2012a, p. 1) gathered for two days to discuss energy and environmental policy options in Japan. The Japanese deliberative poll was carried out by the Center for Deliberative Polling at Keio University led by Professor Yasunori Sone, in collaboration with Professor James Fishkin at the Center for Deliberative Democracy at Stanford. The poll was the first deliberative poll ever to be commissioned by the Japanese government (CDD 2012a, p. 1), and it was making a statement by commissioning the poll that they wished to consult the public on this matter. The background to the poll was the establishment of the Energy and Environmental Committee after the 2011 Great East Japan Earthquake and the Fukushima Dai-ichi disaster, which proposed three options for energy and environmental policy for the year 2030 (CDD 2012b, p. 4). The Committee promoted a national debate around these three options, and the poll was part of the consultation process. The three policy options ranged from zero reliance on nuclear energy to gradual reduction, or keeping the status quo:

- 0 per cent option: We should abandon all nuclear power plants as soon as possible, shutting down all plants by 2030.
- 15 per cent option: We should reduce nuclear power generation gradually (as a result, the ratio of nuclear power generation in total electric power will be about 15 per cent).
- 20–5 per cent option: We should keep nuclear power generation on a lower ratio as compared to the current level (as a result, the ratio of nuclear power generation in total electric power will be about 20–5 per cent). (Briefing material distributed to participants, CDD 2012b: 4)

The main aim of the poll was to 'find a path until 2030' (CDD 2012b, p. 2) by deciding at the end of deliberation which of these options this representative sample of the Japanese public considered most desirable. 6,849 people were contacted and completed a questionnaire, and out of those, 285 people from all over Japan also agreed to take part in the poll.

The main finding of the poll is that the sample public was against reliance on nuclear energy in the future at the beginning of the poll, and that they became even more committed to their original position after the provision of information and deliberation (CDD 2012a, p. 1). Returning to the three policy options posed by the Japanese government, here we look at the changes in attitudes to the 0 per cent, 15 per cent and 20–5 per cent option in turn. In the initial survey (when participants were originally contacted) the participants scored a mean of 6.9 from a scale of 0 to 10 (10 = strongly agreeing with the 0 per cent option), which increased slightly to 7.3 at the end of the two-day deliberation, though the change was not statistically significant (CDD 2012c, p. 2). This means that the small increase cannot be concluded to be a change in attitude which can be generalized to the Japanese population, but shows that the participants came out of the deliberative process continuing to support the 0 per cent policy. As for the 15 per cent and the 20–5 per cent options, there was a statistically significant attitudinal shift towards decreased support. The participants in the initial survey gave a mean of 6.1 in support of the 15 per cent option, which decreased to 5.2 at the exit survey (significant at the .01 level using a dependent samples t-test). Support for the 20–5 per cent option fell from a mean of 4.3 to 3.4 (significant at the .001 level; CDD 2012c, pp. 2–3). In sum, the participants continued to support the zero nuclear energy option and support fell for policy options which opted for a reduced use or a status quo use of nuclear energy.

Looking closer at some of the other attitudinal changes that occurred during the deliberation process, we can see that participants were keen to save more energy at an individual level as well as at an industrial level. When asked whether the Japanese people have already saved enough energy or should save more there was a statistically significant increase in support for more energy saving (CDD 2012c, p. 4). There was also a similar significant increase in support for Japanese industry saving more energy. Alongside a strengthened commitment to move away from nuclear energy and to reduce energy consumption, there was a move away from placing responsibility on the state to the community, and in making positive changes from the nation to the individual. Participants who considered it necessary to 'solve the energy issue as a community, and to buy and use locally' versus considering 'it is efficient to create a large volume of electricity as a nation' changed their views more in favor of *locally* dealing with the issue (CDD 2012c, p. 5). When participants were asked whether 'switching energy policies should be done by the government responsibly' or 'each individual needs to ready themselves for a switch in energy policy and change their thinking', they moved towards individual responsibility and away from the state (CDD 2012c, p. 5). This demonstrates that the poll empowered the participants to think more locally and individually about the future of energy, and to believe more in a bottom-up change in how energy is managed in Japan.

We have discussed the deliberative poll in some detail because the rationale behind deliberative polling is very similar to the ethos behind human security. Both are committed to the empowerment of people by providing information, education and opportunities for deliberation and debate. And both are committed to participation, to the scrutiny of existing social arrangements and ultimately, hopefully, to the possibility of collective action.

Evaluation: the need for people-centeredness

As we have seen, there is a convincing portfolio of evidence that a majority of the Japanese public are opposed or have doubts about nuclear energy. However, while the deliberative poll provided us with valuable information about the informed value-preferences of the Japanese public, it is also already out of date in two important ways. Three months after the poll, Abe won an election victory running on a clear pro-nuclear ticket, and won a resounding victory, in what must be regarded as a high-profile and significant blow to the anti-nuclear movement. As a result, the zero option for nuclear power is off the table in the short term. There will be restarts; it is a question of how many, not whether there will be any.

Whilst being openly pro-nuclear in one sense, however, Abe has adopted a somewhat stealthy strategy in another. He has said that it will take many years to develop a comprehensive energy policy (Kingston 2013a; Squassoni and Gorman 2013), thereby deliberately and cleverly avoiding the larger debate on the energy mix question. METI's December 2013 redraft of Japan's long-term energy plan is consistent with this strategy. The redraft emphasizes the importance of nuclear energy, and identifies nuclear power as an 'important base-load power source that serves as a foundation' for the stability of Japan's energy supply, but it does not commit to percentages (Kyodo 2013). Although he has made a general commitment to nuclear energy, both in the election campaign and since, Abe has focused more on economic issues, another dimension of his 'stealthy' strategy. Whilst acknowledging the significance of the Abe victory, however, we are also entitled to ask how much of a victory it was in terms of support for nuclear power. Exit poll evidence produced by the *Yomiuri Shimbun* suggests that 30 per cent of voters listed the economy as the most important issue for them, with only 13 per cent placing nuclear issues at the top of their list of priorities (Squassoni and Gorman 2013). If most people are more concerned with pocket-book issues, and the perception is that Abe is the only game in town as a solution to these problems and is campaigning almost exclusively on economic issues, then can we really say that this is a crushing defeat for the anti-nuclear movement? Furthermore, it is clear that if we isolate the nuclear issue from the pocket-book issue, opposition to nuclear energy is very strong indeed; another set of exit polls conducted by *Asahi Shimbun* on 16 December 2012 found that 78 per cent of respondents favored either an immediate or gradual move toward a nuclear-free society, much larger than the 15 per cent who opposed such moves (*Asahi Shimbun* 2012).

We can also question the extent to which this was a resounding victory for Abe, as opposed to being primarily a rejection of the DPJ. In an *Asahi Shimbun* survey 81 per cent of respondents answered that the LDP's victory was because of 'disappointment with the DPJ government', compared with only 7 per cent who thought it was due to 'support for the LDP's policies'. Voter turnout was the lowest in Japan's post-war history, at 59.3 per cent ten percentage points lower than in 2009, which is striking given that this was the first election following the triple disaster (Hobson 2012). The elections were a vote against the DPJ and for disenchantment, rather than an endorsement of Abe's nuclear energy strategy in any meaningful sense.

Abe's short-term strategy is clearly to get as many reactors as possible restarted, without begging the larger question on the energy mix in Japan, and hopefully getting so many reactors restarted that nuclear energy will be locked in, and will become a fait accompli. Abe will have some relative success with this strategy (Kingston 2013a). However, precisely because Abe is adopting this 'stealth strategy', there is a probability that the larger energy mix question will remain unaddressed, unresolved and therefore in play at the time of the next elections, in 2016. Much will depend, therefore, on the resilience of the NRA to the substantial pressure being put on it by the Abe administration to authorize restarts. Much, in turn, will also depend on continued opposition from other key players, such as the media, anti-nuclear political parties (particularly New Komeito) and civil society. The media and public opinion can play a key role here. The activities and decisions of the NRA must be forensically scrutinized and publicized by the media, and the public kept extremely well-informed about what is at stake in these technical deliberations that the government would like to keep at a low-profile, 'expert' level. The media, and especially the explicitly anti-nuclear media, must make this a strategic priority for the next two to three years. It is to be hoped that public opinion and dissent can be stimulated as much as is possible. The *Japan Times*, as we saw, appealed to the public to mobilize to prevent restarts, and now it must do all that it can to *limit* the number of restarts, and to present information that makes it easier for the public to be aware of the issues, and to mobilize. Deliberative polling can also potentially play an important role, in terms of educating the Japanese public, generating media attention and providing valuable qualitative information on what the public would think if they had the time and information to make well-informed and considered moral and practical judgments about energy mix questions. Perhaps it is time for a new deliberative poll, reflecting the choices now realistically available to us. Such a deliberative poll could be heavily publicized by the media, and could offer a fresh attempt to empower the Japanese public, and thereby contribute to their human security.

Conclusion: the interrelated nature of threats to human security

The *HDR* identified seven categories of threat to human security, all of which are relevant when considering Fukushima and nuclear power in Japan: economic security, food security, health security, environmental security, personal security, community security and political security (UNDP 1994, pp. 24–33). With regard to economic security, the Abe administration and the nuclear village want to present the nuclear option as the most compelling. From their perspective, nuclear energy is cheaper, the cost of importing additional fossil fuels to cover the nuclear shortfall has been prohibitive and Japan can make large amounts of profit from the export of its nuclear technology and expertise. On the other side of the argument, it can be claimed that there is too much financial risk associated with construction cost overruns, extended shutdowns and the incalculable costs associated with catastrophic nuclear failure that can, as we have seen, be triggered by natural disaster. It is also not clear whether the cost of fossil fuels to cover the nuclear shortfall can be chalked up as part of the argument in favor of nuclear energy, or as part of the financial risk associated with nuclear crisis. Some experts argue that it is necessary to protect the trapped costs of nuclear investment, and that Japan cannot afford not to include nuclear in its energy portfolio if it wants to avoid running massive deficits. Others argue that in the longer term costs associated with renewables can be reduced, and there is considerable scope for the renewable sector to grow as a result of energy market liberalization and government investment.

With regard to political security, the Abe victory is, on the face of it, a significant blow to the anti-nuclear lobby, given that Abe campaigned and won on a pro-nuclear ticket. However, we have argued that the election represented a combination of voter apathy and a desire to punish the DPJ, rather than a meaningful endorsement of the Abe administration, especially beyond its economic policies. Abe campaigned almost exclusively on economic issues, and twice as many people identified economic issues as an election priority. However, it can also be argued that, the election result notwithstanding, the Abe administration is ignoring the substantial and sustained anti-nuclear sentiment that exists within Japan, a strategy which is certainly not consistent with the democratic and human rights of the population, and which can ultimately be argued to be threatening to the electorate's human security. The volume of opinion poll evidence and the conclusion from the deliberative poll can be seen, taken together, as providing robust evidence of a consistent opposition to nuclear energy on behalf of the Japanese public, and as a more credible indication of true considered judgment than the election result.

With regard to environmental security, we must note that renewable energy is the most desirable option, and also that there is a trade-off between the environmental harm caused by fossil-fuel engendered global warming, and the risks associated with reactor failures and nuclear meltdowns. Japan abandoning nuclear power would lead to a significant increase in Japanese fossil-fuel consumption and emissions. But the environmental risks associated with reactor failures and nuclear meltdowns are also potentially catastrophic, and include the spread of radioactivity into the atmosphere and into the food chain. Food security is obviously threatened by such developments.

The downside risk associated with potential nuclear catastrophes also speaks to concerns with health, personal and communal security. In terms of health security, people are of course at risk of direct exposure through the atmosphere, and delayed exposure via the food chain. Food security is obviously also threatened by such developments. Individual workers such as the 'Fukushima Fifty', who were perceived by some to have had a professional and moral responsibility to directly address the meltdown, were at severe risk (see Hobson's chapter in this volume). Forced relocation *away* from the disaster area as a result of nuclear catastrophe produces a range of social insecurities at the interface between community and personal security; these include the social and sexual vulnerabilities of women living in temporary accommodation, and the longer term social risks of 'atomic divorce' and 'dying alone'. Forced relocation *into* the disaster area is also a personal human security threat for those at the margins of society who find themselves working as part of the relief labor force.

In terms of community security, the reactors are placed in poorer regions and areas, to ensure that local populations are grateful for the jobs and investment. It should be noted that many regions that contain nuclear reactors voted strongly in favor of the LDP, and are lobbying hard for reactor restarts. However, one is also entitled to wonder on their behalf whether they are being exploited, and whether nuclear risk is being franchised, at the personal and community level, to poorer regions and local communities. When talking of communal security we can also talk about the general communal security of the Japanese population taken as a whole, in the face of the risks associated with nuclear disaster. We can also consider the communal security of more specific populations which do not host the reactors, but which are profoundly vulnerable to the consequences of their failure, for example, the population of Tokyo, who faced the risk of mass evacuation, with all of the impacts to economic, health, personal and community that such an upheaval would have entailed.

Thinking in terms of the seven categories of human security is useful in that this provides us with a list of human security issues to take into consideration, and also to balance against each other. We can profitably consider threats and risks within each of the categories, and also look at the possibilities for threat multiplication and threat solution across the categories. This in turn informs us about potential trade-offs that may be available or necessary in response to the overall range of threats to human security. In some categories, however, we can produce balanced arguments on both sides of the ledger, as appears to be the case with regard to the economic dimension of human security and the nuclear-mix question. It is here that the four elements of human security can provide further trumping, tie-breaking impetus to the resolution of a debate. We have tried to provide a balanced account of the economic arguments for and against nuclear energy, and have ourselves tentatively come down on the side of the anti-nuclear economic argument. A reasonable case could be made on either side. But if we take seriously the argument for prevention, and consider the full range of human security threats across the seven dimensions of human security, and take into account threats to political, environmental, food, health, personal and community security, it seems clear to us that the costs and risks of nuclear energy outweigh the benefits. To add even more weight to this conclusion, we can also say that the commitment to prevention, which is one of the four elements of human security around which we have structured our discussion, performs a further trumping function, requiring us to place extra negative weight on the downside risk of nuclear failure.

The commitment to people-centeredness, also a key element of the human security approach, further requires us to place additional weight on the enduring and considered views of the people at large. We have argued above that, the election result notwithstanding, the Japanese population is clearly opposed to the nuclear option, and that the deliberative poll is an important part of the evidence leading us to this conclusion. In each of these ways, considering the energy mix question through a human security lens inclines us towards an anti-nuclear position on the Japanese energy mix question, ultimately informed by the twin imperatives of prevention and people-centeredness. It also inclines us to hope and to believe that, given the current 'stealth' strategy of the Abe administration, there is still plenty of scope for a substantial energy mix debate to be a defining and empowering dimension of the next Japanese elections in 2016.

References

Aldrich, D., Platte, J. and Sklarew, J. 2013, 'What's ahead for Abe's energy agenda?' Asia Unbound, *Council on Foreign Relations*, 30 July, http://blogs.cfr.org/asia/2013/07/30/aldrich-platte-and-sklarew-whats-ahead-for-abes-energy-agenda, viewed Dec. 2013.

Asahi Shimbun 2012, 'Pro-nuclear bureaucrats back in the picture under Abe', 29 Dec., http://ajw.asahi.com/article/behind_news/politics/AJ201212290056, viewed Dec. 2013.

Center for Deliberative Democracy 2012a, 'Executive summary: The national deliberative poll in Japan, August 4–5, 2012 on energy and environmental policy options', http://cdd.stanford.edu/polls/japan/2012/jp-energy-executive-summary.pdf, viewed Dec. 2013.

Center for Deliberative Democracy 2012b, 'Briefing material', http://cdd.stanford.edu/docs/2012/jp-energy-policy.pdf, viewed Dec. 2013.

Center for Deliberative Democracy 2012c, 'T1–T3 attitude change', http://cdd.stanford.edu/polls/japan/2012/jp-energy-attitudes-t1-t3.pdf, viewed Dec. 2013.

Center for Deliberative Democracy 2012d, 'T2–T3 attitude change', http://cdd.stanford.edu/polls/japan/2012/jp-energy-attitudes-t2-t3.pdf, viewed Dec. 2013.

Commission on Human Security 2003, *Human Security Now*, New York: CHS, http://isp.unu.edu/research/human-security/files/chs-security-may03.pdf, viewed Dec. 2013.

DeWit, A. and Hobson, C. 2013, 'Abe at Ground Zero: The consequences of inaction at Fukushima Daiichi', *Asia-Pacific Journal*, 11/35(1), 2 Sept.

Dickie, M. 2012, 'Japanese anti-nuclear demonstrations grow', *Washington Post*, 16 July, http://www.washingtonpost.com/world/asia_pacific/japanese-anti-nuclear-demonstrations-grow/2012/07/16/gJQAPXPgoW_story.html, viewed Dec. 2013.

Fackler, M. 2012, 'Japan weighed evacuating Tokyo in nuclear crisis', *New York Times*, 27 Feb., http://www.nytimes.com/2012/02/28/world/asia/japan-considered-tokyo-evacuation-during-the-nuclear-crisis-report-says.html?_r=0, viewed Dec. 2013.

Fishkin, J. S. 1997, *Voice of the People: Public Opinion and Democracy* (new edn), New Haven, CT: Yale University Press.

Fujisaki, M. 2013, 'Japan's growth in renewable energy dims as nuclear strives for comeback', *Asahi Shimbun*, 7 July, http://ajw.asahi.com/article/behind_news/Politics/AJ201307070012, viewed Dec. 2013.

Habermas, J. 1984, *The Theory of Communicative Action*, vol. 1, *Reason and the Rationalisation of Society*, tr. Thomas McCarthy, Cambridge: Polity Press.

Habermas, J. 1987, *The Theory of Communicative Action*, vol. 2, *Lifeworld and System: A Critique of Functionalist Reason*, tr. Thomas McCarthy, Cambridge: Polity Press.

Hobson, C. 2012, 'Wakeup call for Japan's politicians', *Japan Times*, 30 Dec., http://www.japantimes.co.jp/opinion/2012/12/30/commentary/wakeup-call-for-japans-politicians, viewed Dec. 2013.

Japan Times 2013, 'Change Japan's Energy Policy', 17 Sept., http://www.japantimes.co.jp/opinion/2013/09/17/editorials/change-japans-energy-policy, viewed Dec. 2013.

Jiji Press 2013a, 'Almost 60% of public opposes Japan's export of nuclear tech: survey', *Japan Times*, 16 June, http://www.japantimes.co.jp/news/2013/06/16/national/almost-60-of-public-opposes-japans-export-of-nuclear-tech-survey, viewed Dec. 2013.

Jiji Press 2013b, 'Thousands mass for antinuclear rally in Tokyo', *Japan Times* 13 Oct., http://www.japantimes.co.jp/news/2013/10/13/national/thousands-mass-for-antinuclear-rally-in-tokyo, viewed Dec. 2013.

Kingston, J. 2013a, 'Abe's nuclear energy policy and Japan's future', *Asia-Pacific Journal*, 11/34(1), 19 Aug.

Kingston, J. 2013b, 'Tepco's follies, reactor restarts and awkward plutonium stockpiles', *Japan Times*, 31 Aug., http://www.japantimes.co.jp/opinion/2013/08/31/commentary/tepcos-follies-reactor-restarts-and-awkward-plutonium-stockpiles, viewed Dec. 2013.

Kyodo Press 2013, 'Energy plan revised for nuclear emphasis', *Japan Times*, 13 Dec., http://www.japantimes.co.jp/news/2013/12/13/business/energy-plan-revised-for-nuclear-emphasis, viewed Dec. 2013.

Morton, O. 2012, 'The dream that failed', *The Economist*, 10 March, http://www.economist.com/node/21549098, viewed Dec. 2013.

Noyes, T. 2011, 'The incalculable cost of nuclear power', *Guardian*, 3 April, http://www.theguardian.com/commentisfree/cifamerica/2011/apr/03/nuclearpower-japan, viewed Dec. 2013.

Sieg, L. 2013, 'Japan PM Abe's mandate is much smaller than it looks', *Reuters*, 22 July, http://www.reuters.com/article/2013/07/22/us-japan-election-mandate, viewed Dec. 2013.

Soble, J. 2013, 'Koizumi issues strong call for abandonment of nuclear power', *Financial Times*, 12 Nov., http://www.ft.com/intl/cms/s/0/38dd2194-4b86-11e3-8203-00144feabdc0.html#axzz2mqrIk5WT, viewed Dec. 2013.

Squassoni, S. and Gorman, R. 2013, 'Japanese nuclear policy after the 2013 Upper House elections', Center for Strategic and International Studies, 24 July, http://csis.org/publication/japanese-nuclear-policy-after-2013-upper-house-elections, viewed Dec. 2013.

The Economist 2013, 'Electricity in Japan: Power struggle', 21 Sept., http://www.economist.com/news/asia/21586570-shadow-fukushima-worlds-worst-nuclear-disaster-after-chernobyl-hangs-over-japans-energy, viewed Dec. 2013.

Tokyo Shimbun 2012, 'Gempatsu Zero: "Henkou Yochi Nokose" – Kakugi Kettei Kaihi – Beiga Youkyuu', 22 Sept., http://www.tokyo-np.co.jp/article/feature/nucerror/list/CK2012092202100003.html, viewed Dec. 2013.

UNDP 1994, *Human Development Report 1994*, New York: Oxford University Press.

United Press International 2013, '60,000 protest Japan's plan to restart nuclear power plants', 2 June, http://www.upiasia.com/Top-News/2013/06/02/60000-protest-Japans-plan-to-restart-nuclear-power-plants/UPI-34961370197818, viewed Dec. 2013.

World Nuclear Association 2013, 'Nuclear power in Japan', http://www.world-nuclear.org/info/Country-Profiles/Countries-G-N/Japan, viewed Dec. 2013.

Yankelovich, D. 1991, *Coming to Public Judgment: Making Democracy Work in a Complex World*, New York: Syracuse University Press.

Yomiuri Shimbun 2013, 'Japan criticized for lowering goal', 25 Nov., http://the-japan-news.com/news/article/0000822106, viewed Dec. 2013.

11 Towards human security

Climate change and the military role in humanitarian assistance and disaster response

Andrew DeWit

Introduction[1]

This chapter explores how accelerating climate change is amplifying the military role in foreign humanitarian assistance and disaster response (HADR). On the basis of the evidence, I argue that it is imperative to encourage and indeed expand a military-centred HADR collaboration in order to better promote and protect human security. The collaborative frameworks that should be institutionalized are those between the military and civil-society HADR networks (such as NGOs) and among the American and other global militaries (especially the Chinese). I discuss several reasons for these objectives at greater length below. Primary among them are that only military forces have the heavy air- and sea-lift capabilities often required for intervening in increasingly frequent and intense disasters. They also possess the unique capability of providing crucial security operations in the chaos following disasters (McGrady et al. 2010) as well as immediate relief via such means as hospital ships (NRC 2011, p. 5).

There is a certain irony in this, because the military are generally and correctly seen as a threat to human security. The reality of this threat is routinely demonstrated in a number of dimensions. It manifests itself in drone strikes and other military operations that take lives. It is implicit in massive defence spending, discussed below, that impoverishes lives by diverting scarce resources. And it abides in the perennial threat of nuclear war. Yet at the same time, climate security threats appear increasingly collective and existential, requiring ever more frequent and large-scale HADR, where the military act as protectors and providers of human security. This undeniable fact makes it imperative to work with the military. Indeed, this opens the opportunity to help it evolve in a constructive direction, to civilize the military rather than risk further militarizing society.

Recent examples of the constructive HADR option include the massive and very effective 'Operation Tomodachi' military and NGO relief operations in the wake of Japan's 11 March 2011 earthquake and tsunami natural disasters (Samuels 2013). The collaborative security role was highlighted even more strongly by the mid-November 2013 devastation wrought in the Philippines by Typhoon Haiyan. Not only was the storm unprecedented in ferocity, and thus stark evidence of climate-driven mounting security threats, but HADR operations also saw the dispatch of American, Chinese, Japanese and other countries' naval assets. Indeed, these countries competed to collaborate (Spegele 2013), a pertinent fact that illustrates the HADR community's opportunity in otherwise ever more fraught times.

Climate change and HADR

Among international initiatives, HADR is a large but woefully underfunded sector. The demand for HADR is increasing: between 2001 and 2011, 'the annual average number of people affected by natural disasters has risen by 232 per cent, compared to the decade 1990 to 2000' (Poole 2013). 2011's count of 332 natural disasters was fewer than the annual average of 384 for the decade 2001 to 2010. But in 2011 deaths totalled 30,773, with 244.7 million other categories of casualties and a record-setting US$366.1 billion in economic damage (IFRC 2013). This level of economic damage is over twice the volume of overseas aid flows (*NGO Voice* 2013). The United Nations makes a larger claim, arguing that, since the start of the millennium, natural disasters have taken 1.1 million lives, affected 2.7 million others, and cost US$1.3 trillion (Ban 2013). This damage is not distributed equitably across the globe. Geographically, the *Global Climate Risk Index 2013* shows that the countries most affected during 2011 were Thailand, Cambodia, Pakistan, El Salvador and the Philippines (German Watch 2012).

The toll of lives and assets also seems certain to increase significantly, even as politics hamstrings the capacity to respond via public governance. The September 2013 release of the Intergovernmental Panel on Climate Change (IPCC) 5th Assessment Report, warned of mounting risks, and raised the level of certainty that most global warming is man-made from 90 to 95 per cent. Subsequent reports are expected to amplify the emphasis on risks (Doyle 2013). These reports are labelled as alarmist by denialists bent on defining the limits of acceptable debate and further distracting public agency. But of course the facts are entirely to the contrary. The IPCC process omits from its purview such significant feedback effects as methane release from thawing permafrost (Schaefer 2012), and the dramatic increase in 'anthropogenic' forest and bushfires and other factors (Glikson 2013). And the IPCC's focus on limiting warming to 2 degrees Celsius is clearly misconceived in light of climate impacts already evident (Hansen et al. 2013).

In September 2013, the OECD highlighted the enormity of the threat posed by the failure of national governments and their international agencies to prepare. Its survey *Water and Climate Change Adaptation: Policies to Navigate Uncharted Waters* is framed with a concise and cogent argument that '[c]limate change is to a large extent water change. Climate change affects all aspects of the water cycle and water is the main way through which the impacts of climate change will be felt'. The OECD also advises that there is a 'growing recognition that climate change presents a singular challenge for water systems by rendering the historical assumption of stationarity increasingly irrelevant'. The best short definition of 'stationarity' is 'the idea that natural systems fluctuate within an unchanging envelope of variability' (Milly et al. 2008). In the OECD's view, the end of stationarity 'means that a fundamental assumption upon which water management, infrastructure design and planning, and ultimately many economic and resource management decisions are founded will no longer be a reliable basis for future planning and management' (OECD 2013).

The end of stationarity means decisions on critical infrastructures that will be used for decades lack reliable measures for how hardened they should be to contend with water, the biggest element of the climate system. Urban managers and others can have no confidence in future levels of precipitation, storm surges and other pertinent facts as well as how rapidly to try and adapt. The loss of stationarity also means that past investments in roadways, waterworks, energy systems and the like may be vulnerable (DOE 2013). The global community saw a startling display of that possibility when Hurricane Sandy hit New York City in late October 2012, knocking out its power grid and turning parts of its subway system into a sewer (Metcalfe 2012). More recently, in late August 2013, 60 per cent of the Philippines capital Manila was flooded by torrential rains that unleashed more than a month's worth of precipitation in a single day (Hranjski 2013). Much of the urban infrastructure that has been built and is being built could become death-traps, particularly for children and the elderly, in the midst of natural disasters.

A further implication of stationarity's eclipse is that most of humanity is rendered more insecure because most large cities lie on coasts, or near other bodies of water (Stone 2012). The World Bank warns that 13 of the world's 20 largest cities lie on the oceanic coast, and nearly a third of the global population is within 160 km of a coast. It also points out that over 50 per cent of the 7 billion global population now lives in cities (World Bank 2010). Moreover, this share is expected to increase to 60 per cent by 2030 and 70 per cent by 2050, the latter number representing some 6.4 billion people (WHO 2013). Thus, under business as usual, most people will live close to increasingly dangerous shores.

These facts on climate change are neither disputed nor ignored by the military and other institutions (such as the reinsurance industry) that see the world in terms of risk and are compelled to act. The US military's Strategic Environment Research and Development Program (SERDP) reveals that

> [i]n coordination with the efforts of the other federal science providers, SERDP's goal is to ensure DoD [Department of Defense] has the necessary science and tools to support climate change-related vulnerability and impact assessment. A suite of SERDP projects are *developing the methodologies and tools* [emphasis added] needed to assess the physical effects of sea level rise and storm surge and the impacts to mission-essential infrastructure over a broad range of both geophysical settings and extant climate conditions.
>
> (SERDP n.d.)

That point deserves to be underlined. These US federal agencies, with the military at their core, are in the process of constructing analytical mechanisms to appraise and adapt to a multi-faceted phenomenon of unprecedented speed and scale. They have good reason to: the current pace of climate change has recently been authoritatively assessed as 'at least 10 times faster than any climate shift in the past 65 million years' (Carey 2013). Many institutions of civil society and public governance are distracted by the well-funded denialist politics of climate change (Oreskes and Conway 2010). But climate change's increasingly patent and expensive impact on energy, water and other infrastructure has forced military and other institutions sensitive to atmospheric and oceanic signals to respond.

The capacity problem in conventional HADR

The non-military capacity to respond to major disasters is limited, and thus coping with increasingly frequent and large-scale disasters poses a challenge. There are plenty of HADR organizations, tens of thousands of workers and millions of volunteers, but they work with limited financial, technical and material resources. As to the numbers of organizations, the *State of the Humanitarian System 2012* report calculates that 4,400 NGOs are at work in the global humanitarian field (Humanitarian Outcomes 2012). The report laments, however, that the system is dominated by a 'small group of global giants: the UN humanitarian agencies, the International Movement of the Red Cross and Red Crescent, and five international "mega" NGOs, whose combined humanitarian expenditure in 2010 exceeded $2.7 billion'. The major actors in the HADR sector include the International Committee of the Red Cross, which was established in 1863 and employs roughly 12,000 people in 80 countries. The largest humanitarian network is the International Federation of Red Cross and Red Crescent Societies (IFRC), whose 13 million volunteers aid 150 million people in 187 national communities. The IFRC focuses on the four core areas of promoting humanitarian values, disaster response, disaster preparedness and health and community care. In terms of humanitarian workers deployed in the field, the HADR network has itself expanded by 6 per cent per year over the past decade.

The above numbers suggest that, in addition to the escalating climate threat, another incentive for the HADR community to collaborate is shrinking finances. The 2012 global tally of US$17.9 billion in public-private contributions represented a significant decrease from 2011's total of US$19.4 billion and 2010's US$20.2 billion. Among the most significant government-sector cuts in contributions to humanitarian assistance was depression-bound Spain's halving of its funding to US$230 million, Japan's 38 per cent reduction of its contribution to US$606 million, and the American trimming of its contribution by 11 per cent to US$3.8 billion (GHA 2013).

At present, one of the largest international efforts for coordinating the NGOs involved in HADR is the Consortium of British Humanitarian Agencies (CBHA). The CBHA was set up in 2010 with a pilot grant of £8 million from the UK Department for International Development (DFID). The CBHA consists of 18 member NGOs that collaborate on enhancing NGO capacity for HADR. They argue that NGOs do the bulk of humanitarian work but are inadequately funded. They would like to see such dedicated funding as the UN's Central Emergency Response Fund (approved by the UN in 2005, launched in 2006 and allocated US$477.3 million in 2012) made more readily available to NGOs.

In short, the conventional HADR realm includes a lot of players contending for dwindling resources in a context of increasing need. This is a recipe for an inadequately robust response to the already significant and rapidly expanding threat of disasters.

The elephant in the room

The elephant in the room when it comes to HADR capacity is the military, which has ample resources and – often – the will to bolster resilience in the face of climate change. Much of this section will centre on the activities and capacities of the US military, the biggest, best-equipped and most globally distributed of all. The US military composes 41 per cent of global military spending. Indeed, the US military's US$645.7 billion defence budget for 2012 vastly outpaced its nearest rival, China's US$102.4 billion defence budget.

Recall the financial resources available to the international humanitarian system, discussed earlier. The year 2012 saw a total of US$17.9 billion for humanitarian assistance (of which HADR is just one function). The US military's annual budget of over US$600 billion is thus well over 300 times global humanitarian assistance outlays. And total global defence spending in 2012 was US$1.582 trillion. Defence expenditures are thus in a different league altogether when compared to conventional HADR.

In a rational and humane world, this stupendous expenditure on defence and skimping on the humanitarian system would be reversed. Conventional military spending generally undermines human security by exacerbating the risk and stakes of conflict. Accelerating climate change increases the damage inflicted by the water-energy-food nexus (UNESCAP 2013), especially in the context of economic upheaval. Much academic work appears stuck in a world of stationarity, debating on the basis of the past whether and to what degree climate change and such related factors as disease, starvation and the like are drivers of conflict (Theisen et al. 2013). But based on its own experience and analyses, the US military is convinced that onrushing climate change is a 'threat multiplier'. More generally, the US intelligence community is deeply concerned about climate change's role in exacerbating conflict and other forms of instability (Steinbruner et al. 2013).

Given the speed and scale of the climate threat, it would make eminent sense to shift massive resources from military spending to HADR and the humanitarian system as a whole. Such a reallocation of resources is not, apparently, on the political agenda anywhere. Nor would a reduction in military spending, even if it happened, necessarily lead to more spending on enhancing resilience. These are additional reasons that it seems most fruitful to investigate how military resources are being deployed in HADR and how they might be increased. The proactive response of the US military to global warming, evident in its commitment to HADR, suggests that it might be possible to accelerate the evolution of military priorities, through collaboration in the face of twenty-first-century security and humanitarian challenges.

The military role in HADR

The military role in HADR has a very long history, with own-country deployments dating back to ancient times. But it appears that large-scale and foreign-country HADR began in the response to Hurricane Georges and Hurricane Mitch that ravaged the Caribbean, the Gulf of Mexico and Central America in late 1998 (Metcalf et al. 2012, p. 15). At the time, Mitch was the fourth-strongest hurricane ever recorded in the Atlantic (as of this writing, it is now seventh-strongest) and perhaps the most destructive in Central America for the past two centuries. Mitch killed over 10,000 people, left 13,000 missing and destroyed roughly 60 per cent of the infrastructure in Honduras, Nicaragua and Guatemala. While fully 12 countries contributed military support, the US military contribution apparently had the most significant impact. A deployment of 1,377 US military personnel saved at least 700 lives and distributed 1.14 million kg of food, just under 265,000 litres of water, and 42,500 kg of medical supplies during just the initial phase of the relief effort. US military helicopters, fixed-wing aircraft and other heavy equipment played essential roles in these relief operations as well as the enormous reconstruction effort that followed (*Global Security* 2011). In total, the US Department of Defense deployed 7,000 personnel to the afflicted region and spent over US$200 million on relief and reconstruction (Lidy and Kunder 2005).

After Mitch, the US military increasingly recognized such 'complex humanitarian operations' as 'harbingers of future military operations'. The US military community also appears to have been influenced by the assessment of the Rendon Group's CEO John Rendon, a self-described 'information warrior' and certainly a core military propagandist in the 'global war on terror' (GWOT). Rendon described the 'US military operation for [25 December 2004 Indian Ocean] Tsunami relief as the only strategic victory in the GWOT in four years' (Josten 2006). Against this backdrop, the American military was deployed at least 40 times between 2004 and 2010 to support HADR operations (Metcalf et al. 2012, pp. 5, 15). And the US military deployment for HADR in the face of Japan's 3/11 natural and man-made disasters, during 12 March to 4 May 2011, involved 24,000 US service members, 189 aircraft, 24 naval ships and cost US$90 million (Johnston 2012).

In spite of these developments, and the increasing frequency of disasters, many actors within the expanding HADR network, as well as analytical observers of military affairs generally, continue to see the military role in HADR as exceptional. Not a few deem it to be undesirable. Indeed, the so-called Oslo guidelines, agreed to in 1994, provide a 'normative and practical framework' for the military role in HADR that describes foreign-country HADR operations by the military as a 'last resort' (Fischer 2011).

The humanitarian community's concern about the military and HADR

Misgivings about the military role in HADR appear strongly represented among the earnest but under-equipped humanitarian agencies working on the front lines of increasingly numerous disasters. The Oslo guidelines' reluctant view of the military role as a 'last resort' clearly imbues much thinking. Not a few HADR agencies and scholars are outright hostile to the military role, their opposition often expressed in terms of fear of being manipulated or caught in the crossfire.

Surveys of the humanitarian literature itself do not elicit empirical evidence of harm to HADR groups by military involvement (Metcalf et al. 2012, p. 10). Yet the military role in HADR may indeed be influenced by political goals that extend well beyond the aim of getting food, water and shelter to those in peril. Recall for example that a prominent feature of the past decade was the costly 'GWOT' geostrategic initiative that Rendon both helped bring into being and criticized. NGOs in the HADR sphere are in the main animated by an ethic that stresses, first and foremost, doing no harm to those they wish to help. Given the way the military has been used in the past (Ferreiro 2012), it is not at all difficult to sympathize with the above concerns. But the US military's training in HADR, such as at its Hawaii-based Center for Excellence in Disaster Management and Humanitarian Assistance, is based on the same humanitarian code (Moler 2013). This fact provides a moral lever in NGO–military collaboration.

Moreover, the merits and demerits of working with the military surely need to be re-examined in light of clear thinking on the HADR implications of climate change. On this point, the military appears to be far ahead of the HADR NGOs. As we shall see presently, military think tanks and other institutional resources have been paying close attention to climate change for some time. By contrast, and in spite of the accumulation of evidence linking climate change to more frequent and intense natural disasters, it would appear that NGOs and other groups at work in the field on HADR only began to talk to climate scientists in early 2013. In mid-May of 2013 the HADR NGOs engaged – through the auspices of the Harvard University Center for the Environment – formally with climate scientists. Harvard University public health expert Jennifer Leaning, a specialist in disaster response and forced migration, was instrumental 'in opening a dialogue' between the HADR NGOs and the climate scientists. Her comments on the event reveal that the humanitarian community was in large measure frustrated. The NGOs and others wanted 'to know certain kinds of qualitative assessments with a real amount of scientific certainty so that we can plan for areas that will be more likely to be hit by drought, areas that are going to have serious sea level rise, and areas that are going to be most at risk for very heavy storms' (Toomey 2013). The dismay of the HADR NGOs was due to the climate scientists' inability to forecast the micro-level fallout of climate change, for example, to pinpoint where a flood will take place so that they can adequately prepare.

It is quite understandable that the civil-society HADR community want to be able to anticipate events, and preposition scarce human and material resources. They do not possess plentiful financial and logistical resources, and hence lack the means to respond quickly and massively, drawing on prepositioned stocks. But to be blunt, physics and chemistry do not cooperate by sending advance notice of large-scale disasters. So to the extent that there is in fact accelerating climate change under way, it follows that the humanitarian system will have to rely more frequently on military sea-lift, air-lift and other capacity in order to mount adequate responses.

This is not a novel argument: the same conclusion was reached in 2011 by the US National Research Council of the National Academies, in its study *The National Security Implications of Climate Change for US Naval Forces*. Among other recommendations, the report argued for greater military investment in HADR-capable vessels and infrastructure. Of particular concern was the emphasis on maintaining hospital ship capacity as well as hedging 'against climate change impacts through planning for modifications of the existing force structure as climate change requirements become clearer'. In addition, the study determined that confronting climate change required a risk-based approach, and strongly indicated the need to be proactive. In other words, it made the case for mobilizing resources on the basis of overwhelming evidence rather than waiting for iron-clad scientific certainty which does not exist.

The National Research Council report also addressed the collaborative dimension of responding to HADR in particular, as well as the ravages of climate change more generally. It declared that '[g]iven the scope and scale of potential climate change contingencies, and the projected global climate change vulnerabilities, the United States and its naval forces will want to cooperate with allies, non-allies, and private organizations in both anticipating and responding to global climate change and geographic hot spots'. The report emphasized a broad range of collaboration, and specifically included China, Russia and NGOs in its list of 'nontraditional partners' with which the US military should collaborate (NRC 2011, pp. 11–12).

The military lead on HADR

The US military's attention to climate change and its HADR implications stretches back at least to the mid-2000s. In order to assess the threat of climate change, the US military's Center for Naval Analysis, a US federal-government funded research and development centre for the Navy and the Marine Corps, convened a Military Advisory Board (MAB) of 11 retired three-star and four-star admirals and generals in 2006. The MAB assessment was released the next year, in a landmark report *National Security and the Threat of Climate Change*' (CNA 2007). The report depicted climate change as a 'threat multiplier for instability in some of the most volatile regions of the world'. It also cautioned that climate change had to be 'fully integrated into national security and national defense strategies'. But even more than that, the report urged that the US take 'a stronger national and international role to help stabilize climate change'. The report also stressed the importance of 'global partnerships that help less developed nations build the capacity and resilience to better manage climate impacts' (CNA 2007).

This report, too, acknowledged the lack of 100 per cent scientific certainty. It declared that scientific certainty was not likely ever to be achieved, but that the risks were so potent that action was imperative. Retired General Gordon R. Sullivan, chairman of CNA's MAB and the US Army's former chief of staff, wrote that:

> the trend line is very clear. We never have 100 percent certainty. We never have it. If you wait until you have 100 percent certainty, something bad is going to happen on the battlefield. That's something we know. You have to act with incomplete information. You have to act based on the trend line. You have to act on your intuition sometimes.
>
> (Cimons 2013)

The CNA followed up with a May 2009 analysis of climate change and conventional energy, titled *Powering America's Defense: Energy and the Risks to National Security*. This study offered a detailed investigation of the dangers posed by climate change, the reliance on conventional energy and other elements of 'a series of converging risks'. It stressed that climate change 'has the potential to create sustained natural and humanitarian disasters on a scale far beyond those we see today'. The analysis also recommended that priority be given to helping nations 'build the capacity and resiliency to better cope with the effects of climate change. Doing so now can help avert humanitarian disasters later' (CNA 2009, pp. 6–7).

This long background of attention to climate change within military and related circles sparked, for example, considerable activism within the US Navy. This should come as no surprise, since the Navy works in the medium – water – that most directly manifests climate change. In May 2009, the Chief of Naval Operations created Task Force Climate Change (TFCC). The TFCC was given a broad mandate of making recommendations to the 'Navy leadership regarding policy, investment, and action, and to lead public discussion on this serious issue'. The Navy also put serious organizational heft behind the TFCC by backing it up with a flag-level Executive Steering Committee (ESC) and making the Oceanographer of the Navy Director of the TFCC and chair of the ESC. The ESC includes officials from a broad range of organizations, not just a host of relevant Navy offices but also the Military Sealift Command Headquarters, the Naval War College, the US Coast Guard and the National Oceanic and Atmospheric Administration (Freeman 2009).

In light of the above, it is to be lamented – and is in fact rather mystifying – that there is not yet institutionalized collaboration, let alone significant contact, between the organized ranks of HADR NGOs and the US Navy's institutional and human resources focused on climate change.

Use the opportunity or risk losing it to denialists

We have examined several practical reasons for HADR actors to collaborate with the military on HADR. A further reason is eminently political, and centres on the risk that the military's scientific, strategic openness to critical thinking on climate change and HADR could shift to a more ideological and tactical orientation. The latter is already evident among most conservative analysts. The conservative position on the military role in HADR per se is diverse, with a focus on prospective gains from showing the flag versus the pecuniary and other costs of getting involved. But virtually all American conservative analysts are deeply troubled by the US military's attention to climate change and its willingness to collaborate with 'non-traditional partners'.

One noteworthy example of this opposition was seen in the wake of a 9 March 2013 *Boston Globe* article concerning remarks by Admiral Samuel Locklear. At the time, Locklear was Commander, US Pacific Command, in charge of the Asia-Pacific region and at the centre of the US 'pivot' to the Asia-Pacific. Locklear had been meeting with scholars from Harvard and Tufts Universities, and – perhaps retaining the spirit of open discourse – told the reporter that climate change is core to strategic concerns. He elaborated by pointing out that his 400,000-person command is 'working with Asian nations to stockpile supplies in strategic locations' and is reaching out to 'other armed forces in the region', including China and India (Bender 2013). An angry retort soon emerged from *The National Interest*, a flagship publication among conservative think tanks. The author was Peter Brookes, Heritage Foundation senior fellow and former Deputy Assistant Secretary of Defense for Asian and Pacific Affairs. Citing Locklear's comment that his listeners were sometimes surprised when he describes climate change 'as the biggest threat in the Pacific', Brookes argued that

> [m]any would be surprised – or even shocked – to hear our senior warfighter in the Pacific say that. It's likely that his listeners would expect him to talk about nuclear North Korea or China's military build-up, cyber or space warfare or even the ongoing sovereignty disputes in the East and South China Seas, which involves some of our allies and friends.

Brookes evinced little sympathy with the military's stress on HADR and collaboration in the face of the twenty-first century's pre-eminent challenge to all nations' security. Instead, Brooks derided Locklear as simply 'reading from the national-security gospel according to Team Obama, which has highlighted climate change as an emerging security threat. Obama national-security team members – past and present – seem to have the climate-change threat encoded in their DNA, no doubt to the delight of fawning environmentalists'. Brookes tracks the military's concern to the *2010 Quadrennial Defense Review*, when the Pentagon argued that '[w]hile climate change alone does not cause conflict, it may act as an accelerant of instability or conflict, placing a burden to respond on civilian institutions and militaries around the world' (Brookes 2013).

Brookes's argument is of course wrong both about the climate threat as well as the roots of the military's activism. As noted earlier, the military study of climate change and HADR dates back long before the 2009 start of the Obama administration. One would think that a former Deputy Assistant Secretary of Defense would know that. Indeed, in tandem with the US military's critical inquiries into climate and HADR, it began building on George W. Bush-era renewable and efficiency targets. The military introduced even more robust targets for renewable energy as well as energy efficiency, depicting these initiatives as 'force multipliers' (Kats and Seal 2012).

Moreover, US Congressional Democratic and Republican opposition to explicit climate change targets as well as renewable energy goals then saw the military increasingly become the core of a green industrial-policy effort that binds it to other federal agencies. The military's cooperation is especially close with the Department of Energy, among federal agencies, and it centres on renewable energy, efficiency and such resilient technology as microgrids. The collaboration also extends to a host of state governments, centres of innovation (e.g. Sandia Labs and the National Renewable Energy Laboratory), as well as such large firms as General Electric and Boeing and a growing profusion of SMEs throughout the US and globally (DeWit 2013).

In a further lesson for HADR NGOs, not to mention other observers, the US military is collaborating especially closely with such state and local governments as Hawaii and California. Some states are explicitly using the greening of military bases, and the pivot to Asia, to aid in their own swing to green. The California Energy Commission is, for one, specifically aimed at accelerating deployment of renewables, efficiency and ICT mechanisms across military bases and in order to link those military bases statewide, followed later by nationwide rollout (California Energy Commission 2013).

But this highly politicized background to military initiatives on climate change and energy crises has led to a form of cognitive dissonance among many conservative analysts. They tend to laud the military role per se, but not what the actually existing military is doing in the face of energy and climate crises. Many conservative scholars and activists are only willing to acknowledge climate change if nuclear and large-scale geoengineering are advanced as solutions (Hamilton 2013). Almost all see national security in twentieth-century terms of state-centred international relations, with the military focused only on war-fighting. They tend to be dismissive of efforts to expand collaboration with China, as was seen in August 2013 when US and Chinese military vessels landed helicopters 'on each other's ships for the first time' (Simoes 2013) and in mid-November 2013 via US-China HADR drills near Hawaii (Tiezzi 2013). It bears thinking about how much worse our already perilous collective security threat would be were this faux-realist perspective to become predominant in shaping the US military's response to climate change, its role in HADR, its approach to energy and its reaction to the steady rise of China.

Full-spectrum cognitive dissonance

A related cognitive dissonance is shared by many observers on the other side of the political spectrum. Not a few progressive intellectuals and activists also appear to believe that stationarity still reigns, and are thus also bereft of a good analysis of the US military's efforts to confront climate and energy risks. This lack of attention is no doubt one major reason the HADR NGOs have so little interaction with the military. There are plenty of examples of tendentious progressive academic treatments of the military to choose from. Without suggesting it is particularly egregious, consider the September 2013 article by Mattea Kramer and Miriam Pemberton (Kramer and Pemberton 2013), co-authors of *A People's Guide to the*

Federal Budget. They ask 'Can we have a Military-Green Energy Complex Instead, Please?' They correctly point out that:

> some savvy and forward-looking outfits in the military sector have already begun converting their know-how into green-tech manufacturing. Take, for instance, Bath Iron Works, the largest employer in Maine. For several decades, the company has gotten most of its revenue from building and maintaining destroyers for the Navy. Now, however, it has joined an initiative to develop deep-water, offshore wind power, with the goal of making Maine the leading state in the nation in such technology and the production systems that go with it. Oregon Iron Works has similarly built its expertise by fulfilling contracts for the U.S. military. Now, it's diversifying into renewable energy, noting proudly that its 'history of innovation in the marine industry gives [it] the ability to produce high-quality, cost effective Wave and Tidal energy devices'.

The problem is that this account is akin to describing a few trees while missing a very large forest. These particular firms are almost certainly turning to renewables because of exposure to the military's aims in green energy and efficiency. The US Navy is, for example, committed to supplying 50 per cent of its energy needs via renewable energy by 2020. Indeed, one of the above companies, the Oregon Iron Works, had direct involvement in the concrete programmes to achieve the Navy's renewable-energy goals (Casey 2012). Just as solar-powered radio and lanterns were of great use in the Philippines after Typhoon Haiyan (Lallanilla 2013; McDonell 2013), much of what the military innovates for war-fighting is useful in the disrupted environment of HADR. Sustainable energy is the key to mitigation as well as adaptation and human security in the immediate and long-term aftermath of disaster. Our public debate should be fully cognizant of these facts and already work at exploring mechanisms for renovating the military towards better promoting human security.

Military HADR: means and ends

A powerful current in the US military is clearly acting productively in the face of an unprecedented and collective crisis that could upend the Oslo guidelines and other common-sense ideas on HADR. If the evidence for climate change is as dire as our earlier review suggests, the US military itself probably has no real option but to collaborate even more with the HADR NGOs and other non-traditional partners in the future.

Seen from a world of stationarity, this argument might seem at least an exaggeration, if not outright nonsense. The US military comprises 41 per cent of global spending and the only force with global reach. The keystone of its enormous capacity for power projection is the 10 Nimitz-class aircraft carriers, costing US$14 billion each. These carriers individually displace over 100,000 tons, and their capacity to perform 150 strike missions at once and roughly 700 sorties a day is simply unmatched. Aircraft carriers are capable of moving at over 30 knots (over 55 km per hour), driven by very expensive nuclear propulsion that does not need refuelling. Yet though the carriers seem almost invulnerable for some, they are at an inflection point in terms of their role. They are threatened by a host of much cheaper weapons systems against which they appear to have no effective defence. These systems include conventional submarines, mines and ballistic missiles (Hendrix 2013; Wheeler 2012). The carriers are also ill-suited to fighting in the shallow littoral waters where naval conflict is increasingly expected to be focused. The pace of technological change, coupled with increasing geopolitical instability, thus threatens to render the carriers a vulnerable extravagance.

Concerning the carriers, the US military seeks to evolve a more robust war-fighting role by equipping them with laser cannon, drones and other revolutionary technological advances (Osborn 2013; Vergakis 2013). Other emergent technologies would turn the carriers into 'floating factories'. This vision would see the navy deploy 3-D printing on its carriers and other vessels in order to print out necessary parts, ammunition and the like as well as be able to use resources distributed in the seawater as materials for manufacturing (Llenza 2013). Clearly some of this is still far over the horizon, especially the capacity to print out foodstuffs and manufactured items (including shelters, sanitation facilities and the like) using 3-D printers and with 'ink' derived from substances dispersed in the immediate environment, whether that be seawater or biomass onshore. But several of these technologies are already being deployed.

Yet this race between offensive and defensive technology seems likely to be determined by two salient factors. One is the decreasing cost and increasing pervasiveness of portable threats to such large-scale weapons systems. This problem is likely only to worsen with technological advance, in the context of worsening climate-driven instability. Both trajectories imply increased threats and the willingness to use them. The enemy could just as readily be a well-funded band of pirates off the Somali coast as the geostrategic challenger China. The second factor is climate change itself, which potentially trumps everything else. In the face of climate change, especially the rapid climate change that might be under way (Hansen et al. 2013), the carrier as weapon-system represents the generalized obsolescence of conventional military force and the twentieth-century concept of 'national defence' itself. As the loss of stationarity accelerates, all ideas and technologies need to evolve rapidly.

The Nimitz-class carriers' capacity to deliver 700 sorties in a day, and the command and control capacity that coordinates them, can be used to deliver enormous destruction *or* to carry huge volumes of food, water, medical supplies and other essentials to victims of disaster (Gordon et al. 2006). Their desalination plants alone are able to produce 1.5 million litres of freshwater per day. And their physical size can help to deliver large quantities of supplies, reducing the costs for HADR NGOs and other agencies that otherwise have to rely on 'expensive airfreight' (Kovacs and Tatham 2007).

Nimitz-class carriers may not be the most efficient means to deliver HADR (Greenfield and Ingram 2011). But it was the Nimitz-class USS *George Washington* that composed the core of US naval assets dispatched to the Philippines in the wake of Typhoon Haiyan (Sevastopulo 2013). Similarly, the USS *Ronald Reagan,* again a Nimitz-class carrier, was the focal part of the seabased mode of HADR operations off the coast of Japan following the natural and man-made disasters in March 2011. Indeed, as US Naval Commander Elton C. Parker points out, 'arguably the four most strategic uses of naval power in recent times have all been deployments of an aircraft carrier in support of humanitarian assistance and disaster relief (HADR) operations in Indonesia, Pakistan, Haiti, and Japan' (Parker 2013). These vessels are already built and not likely to be scrapped. Given that presence, and their clearly demonstrated utility in HADR operations for human security, they make a valuable instrument both for illustrating the need to evolve as well as a vehicle for doing that.

Conclusion: bolstering security

It seems unlikely that the military, especially the US military, is going to decline significantly in size. Within the United States, at least, the military is – and long has been – the most respected of all institutions (Gallup 2013). Moreover, geopolitics and the global economy appear to be moving in increasingly unstable and unsustainable directions. The drivers are manifold, and include the knock-on effects from increasing natural disasters, the rise of China as a challenger for hegemony within the Asia-Pacific, the erosion of a resource-intensive model of economic development, the resource scarcity highlighted by Michael Klare (2012) and others who write about 'extreme energy', and the global food crisis increasingly warned of by Lester Brown and the Earth Institute (Brown 2012) as well as other global organizations. All of these trends seem more likely to prop up, if not increase, military spending.

These potential drivers of increased military spending will surely not weaken without deliberate collective action, particularly in the face of the climate change that exacerbates them all. This increasingly unstable context suggests it is advisable to take a long, hard look at what the military is doing. There are powerful currents within it that are responding productively to human insecurity. The loss of stationarity offers few footholds for building resilience, but this may be one. It seems imperative to work towards evolving the military into an institution focused on HADR and human security, rather than one that is focused on twentieth-century notions of national defence with only a side interest in HADR.

Moreover, global military forces are expanding their already vast menu of HADR-applicable technologies. They are adding robust and distributed communications (DARPA 2013), renewable energy generation (Irwin 2013), power storage (Pentland 2013), 3-D printing (Burke 2013) and other enormously useful elements of a twenty-first-century resilience. This capacity-building is particularly notable in the American military's industrial-policy effort to bolster the climate and energy resilience of their 700 conventional military bases and other assets (DeWit 2013). Thus military forces increasingly possess the means needed for prompt and effective HADR intervention as well as long-term rehabilitation. The quick-response capability perforce makes them part of any equation for adapting to the ravages of climate change. So too does their increasing innovation of the means to rebuild lives and communities, setting them on more sustainable and resilient foundations. The latter can alleviate the risk of protracted human insecurity leading to conflict, disease and other costly consequences.

Much as academe urgently needs interdisciplinary work to cope with the challenges posed by climate change, agencies on the front lines of HADR must find ways to break down their silos. There have been attempts in this direction, such as the March 2012 'symposium of cooperation', organized by the US Department of Defense and various NGOs to forge ties among the US Military Health System and NGOs (Pueschel 2012). This yet sparse contact resembles the only recent start of a dialogue by HADR NGOs with climate scientists. As we saw earlier, that contact required a group of specialists to organize an event dedicated to that purpose. Whatever the reasons, surely the gravity of climate change makes a deliberate, institutionalized dialogue between the US military and the HADR NGOs imperative, with the aim of broadening that to include other 'nontraditional partners' of the US military. This kind of collaboration could help cement together the still nascent HADR cooperation between China and the US, potentially drawing in the Japanese and other regional militaries.

This argument is not a camouflaged paean to the US or any other military. It is, rather, a recognition of reality as well as a warning: military forces need to be further incentivized to collaborate on HADR, as much and as rapidly as possible, in order to steer them away from the mounting risk of direct confrontation. The urgency of military cooperation has been a salient fact since the advent of atomic weapons. But it has perhaps never been as exigent as in the present, an era wherein military technologies for war-fighting are ever more distributed, destructive and resilient, while geopolitics itself is increasingly marked by political, economic and environmental instability. In this context, HADR collaboration is in itself an essential form of promoting both national security *and* human security.

Note

1 The author would like to acknowledge Mark Selden's very useful comments on an earlier draft as well as the editors' painstaking work on this version, with the caveat that any remaining errors are my own. This research is also funded by a grant from the Japan Society for the Promotion of Science.

References

Ban, K-M. 2013, 'Resilient design for sustainable urbanization', United Nations Secretary-General's remarks at World Habitat Day event, New York, 4 Oct., press release.

Bender, B. 2013, 'Chief of US Pacific forces calls climate change biggest worry', *Boston Globe*, 9 March.

Brookes, P. 2013, 'Is the climate the biggest threat?', *National Interest*, 21 March.

Brown, L. 2012, *Full Planet, Empty Plates: The New Geopolitics of Food Scarcity*, New York: Earth Policy Institute, W.W. Norton & Co.

Burke, M. 2013, 'From brains to planes, 3-D printing opens world of possibilities for military', *Stars and Stripes*, 20 July.

California Energy Commission 2013, 'Energy Commission awards $1.7 million for military microgrid project', press release, 8 July.

Carey, B. 2013, 'Climate change on pace to occur 10 times faster than any change recorded in past 65 million years, Stanford scientists say', *Stanford Report*, 1 Aug.

Casey, T. 2012, 'Navy helps ocean power technologies develop wave power', *Clean Technica*, 24 June.

Cimons, M. 2013, 'U.S. military prepares for global unrest amid climate fears' (Op-Ed), *Climate Nexus*, 12 July.

CNA 2007, *National Security and the Threat of Climate Change*, Alexandria, VA: Center for Naval Analysis, April.

CNA 2009, *Powering America's Defense: Energy and the Risks to National Security*, Alexandria, VA: Center for Naval Analysis, May.

DARPA 2013, 'Creating a secure, private internet and cloud at the tactical edge', press release, 21 Aug.

DeWit, A. 2013, 'The US military, green energy, and the SPIDERS at Pearl Harbor', *Asia-Pacific Journal*, 11/9(5), 4 March.

DOE 2013, *U.S. Energy Sector Vulnerabilities to Climate Change and Extreme Weather*, Washington, DC: US Department of Energy, DOE/PI-0013, July.

Doyle, A. 2013, 'UN draft stresses risk of global warming, from economy to health', *Reuters*, 8 Nov.

Ferreiro, M. 2012, 'Blurring of lines in complex emergencies: Consequences for the humanitarian community', *Journal of Humanitarian Assistance*, 24 Dec.

Fischer, E. 2011, 'Disaster response: The role of a humanitarian military', Army-Technology.com, 26 July.

Freeman, B. 2009, 'Navy Task Force assesses changing climate', US Department of Defense, American Forces Press Service, 31 July.

Gallup 2013, 'Confidence in institutions', 1–4 June.

German Watch 2012, *Global Climate Risk Index 2013*, Bonn: German Watch.

GHA 2013, *Global Humanitarian Assistance Report 2013*, Bristol: Global Humanitarian Assistance, July, http://www.globalhumanitarianassistance.org/wp-content/uploads/2013/07/GHA-Report-20131.pdf, viewed Dec. 2013.

Glikson, A. 2013, 'IPCC climate trends: Blueprints for tipping points in Earth's climate', *The Conversation*, 29 Sept.

Global Security 2011, 'Operation Fuerte Apoyo (Strong Support), Hurricane Mitch, Joint Task Forces Bravo, Joint Task Forces Eagle', 7 May.

Gordon, J. IV, Wilson, P. A., Birkler, J., Boraz, S. and Lee, G. T. 2006, *Leveraging America's Aircraft Carrier Capabilities: Exploring New Combat and Noncombat Roles for the U.S. Carrier Fleet*, Santa Monica, CA: RAND Corporation Monograph.

Greenfield, C. M. and Ingram, C. A. 2011, 'An analysis of U.S. Navy humanitarian assistance and disaster relief operations', Naval Postgraduate School, June.

Hamilton, C. 2013, *Earthmasters: The Dawn of the Age of Climate Engineering*, New Haven, CT: Yale University Press.

Hansen, J., et al. 2013, 'Assessing "dangerous climate change": Required reduction of carbon emissions to protect young people, future generations and nature', *PLOS ONE*, 3 Dec.

Heeley, L. 2013, 'U.S. defense spending vs. global defense spending', The Center for Arms Control and Non-Proliferation, April 24, http://armscontrolcenter.org/issues/securityspending/articles/2012_topline_global_defense_spending/index.html#contact, viewed 16 March 2014.

Hendrix, H. J. 2013, 'At what cost a carrier?', *Center for a New American Security Disruptive Defense Papers*, March.

Hranjski, H. 2013, 'Floods cover 60 percent of Metro Manila', *PhilStar*, 20 Aug., http://www.philstar.com/headlines/2013/08/20/1112451/floods-cover-60-percent-metro-manila, viewed Dec. 2013.

Humanitarian Outcomes 2012, *The State of the Humanitarian System 2012*, London ALNAP, Overseas Development Institute.

IFRC 2013, 'Online Certificate Course in Disaster Management', International Federation of Red Cross and Red Crescent Societies, http://www.ifrc.org/en/get-involved/learning/opportunities/online-certificate-course-on-disaster-management, viewed Dec. 2013.

IPCC 2013, *Climate Change 2013: The Physical Science Basis*, Working Group I Contribution to the Fifth Assessment Report of the Intergovernmental Panel on Climate Change, New York: Cambridge University Press.

Irwin, C. 2013, 'Using military muscle to advance energy solutions', *Breaking Energy*, 6 Sept.

Johnston, E. 2012, 'Operation Tomodachi a huge success, but was it a one-off?', *Japan Times*, 3 March.

Josten, R. J. 2006, 'Strategic Communication: Key Enabler for Elements of National Power', IO Sphere, Summer, http://www.au.af.mil/info-ops/iosphere/iosphere_summer06_josten.pdf, viewed 16 March 2014.

Kats, G. and Seal, A. 2012, 'Virtual storage: A force multiplier for the US military', *Green Biz*, 29 Nov.

Klare, M. 2012, 'The new "golden age of oil" that wasn't: Forecasts of abundance collide with planetary realities', *Tom Dispatch*, 4 Oct.

Kovacs, G. and Tatham, P. 2007, 'An initial investigation into the application of the military sea-basing concept to the provision of immediate relief in a rapid onset disaster', paper presented at POMS 18th Annual Conference, Dallas, TX, 4–7 May.

Kramer, M. and Pemberton, M. 2013, 'Beating swords into solar panels: Re-purposing America's war machine', *Tom Dispatch*, 19 Sept.

Lallanilla, M. 2013, 'Typhoon Haiyan aftermath: How technology can help', *Live Science*, 13 Nov.

Lidy, M.A. and Kunder, J. 2005, 'Large-scale humanitarian assistance', in *Handbook on the Analysis of Smaller-Scale Contingency Operations in Long Term Defence Planning*, Report prepared by the RTO Studies, Analysis and Simulation Panel, NATO,http://citeseerx.ist.psu.edu/viewdoc/download?doi=10.1.1.215.947&rep=rep1&type=pdf, viewed 16 March 2014.

Llenza, M. 2013, 'Print when ready, Gridly', *Armed Forces Journal*, May.

McDonell, S. 2013, 'Typhoon Haiyan: Solar-powered radio helps survivors in Philippines', *ABC News* (Australia), 16 Nov.

McGrady, E. D., Kingsley, M. and Stewart, J. 2010, *Climate Change: Potential Effects on Demands for US Military Humanitarian Assistance and Disaster Response*, Alexandria, VA: CAN (Center for Naval Analysis), Nov.

Metcalf, V., Mason, S. and Gordon, S. 2012, *Trends and Challenges in Humanitarian Civil-Military Coordination: A Review of the Literature*, PLACE: Humanitarian Policy Group Working Paper, May.

Metcalfe, J. 2012, 'Newly unflooded New York subway still looking pretty horrible', *Atlantic Cities*, 12 Nov. http://www.theatlanticcities.com/commute/2012/11/newly-un-flooded-nyt-subway-still-looks-pretty-horrible/3875/ viewed March 2014.

Milly, P. D. C., Betancourt, J., Falkenmark, M., Hirsch, R. M., Kundzewicz, Z. W., Lettenmaier, D. P. and Stouffer, R. J. 2008, 'Stationarity is dead: Whither water management?', *Science*, 1 Feb.

Moler, C. 2013, 'Humanitarian assistance response training: It is all in the details', United States Pacific Command, 12 Aug.

NGO Voice 2013, *Funding Disaster Risk Reduction*, Brussels: NGO Voice Disaster Risk Reduction Series.

National Research Council (NRC) 2011, *The National Security Implications of Climate Change for US Naval Forces*, Washington, DC: National Academies Press.

OECD 2013, *Water and Climate Change Adaptation: Policies to Navigate Uncharted Waters*, Paris: Organisation for Economic Development and Cooperation, 2 Sept.

Oreskes, N. and Conway, E. 2010, *Merchants of Doubt*, London: Bloomsbury Press.

Osborn, K. 2013, 'Future carriers built to carry drone fleets', *DefenseTech*, 19 July.

Parker, E. C. III. 2013, 'Aircraft carriers and what comes next', 10 Dec. blog post, http://warontherocks.com, viewed Dec. 2013,

Pentland, W. 2013, 'Why the US military is falling in love with electric vehicles', *Forbes*, 8 Sept.

Poole, L. 2013, 'Counting the cost of humanitarian aid delivered through the military', *Global Humanitarian Assistance Briefing*, March.

Pueschel, M. 2012, 'Navy/NGO symposium emphasizes interagency approach to international health missions', *Force Health Protection and Readiness Archived News*, 25 April.

Samuels, R. J. 2013, *3.11: Disaster and Change in Japan*, Ithaca, NY: Cornell University Press.

Schaefer, K. 2012, 'Policy implications of warming permafrost', UNEP, Nov., http://www.unep.org/pdf/permafrost.pdf, viewed Dec. 2013.

Sevastopulo, D. 2013, 'Typhoon Haiyan: US aircraft carrier joins Philippines aid effort', *Financial Times*, 14 Nov.

SERDP n.d., 'Climate change and impacts of sea-level rise', Strategic Environment Research and Development Program, http://www.serdp.org/Featured-Initiatives/Climate-Change-and-Impacts-of-Sea-Level-Rise, viewed Dec. 2013,

Simoes, H. 2013, 'US Navy seeks more cooperation with China in counter-piracy exercise', *Stars and Stripes*, 26 Aug.

Spegele, B. 2013, 'Chinese military sending relief ship to Philippines', *Wall Street Journal*, 20 Nov.

Steinbruner, J., Stern, P. and Husbands, J. 2013, *Climate and Social Stress: Implications for Security Analysis*, Washington, DC: National Academies Press.

Stone, B. Jr. 2012, *The City and the Coming Climate*, Cambridge: Cambridge University Press.

Theisen, O. M., Gleditsch, N. G. and Guhang, H. 2013, 'Is climate change a driver of armed conflict?', *Climate Change*, 117(3), pp. 613–625.

Tiezzi, S. 2013, 'Can humanitarian drills rescue the US–China relationship?', *The Diplomat*, 20 Nov.

Toomey, D. 2013, 'Scientists and aid experts plan for a warmer future', *Yale Environment 360*, 15 Aug.

UNESCAP 2013, *The Status of the Water-Food-Energy Security Nexus in Asia and the Pacific Region*, Position Paper, 15 March, Bangkok: United Nations Economic and Social Commission for Asia and the Pacific.

Vergakis, B. 2013, 'Navy christens next generation of aircraft carrier', Associated Press, 9 Nov.

Wheeler, W. 2012, 'More than the navy's numbers could be sinking', *Time*, 4 Dec.

WHO 2013, 'Urban population growth', Geneva: WHO, http://www.who.int/gho/urban_health/situation_trends/urban_population_growth_text/en, viewed Dec. 2013.

World Bank 2010, *Climate Risks and Adaptation in Asian Coastal Megacities: A Synthesis Report*, Washington, DC: World Bank, Sept.

12 Life after the triple disaster

Human security and the future

Christopher Hobson

A core argument of this book is that the 2011 earthquake, tsunami and nuclear accident clearly demonstrated that the human security approach is relevant to Japan. This has been illustrated in detail by the various contributors. It is important to emphasize that the triple disaster exposed and greatly exacerbated this reality, but it did not create it: serious human insecurities existed in Japan before 11 March 2011. Indeed, these pre-existing patterns of vulnerability were significant factors in determining the way the disaster unfolded. To take one noteworthy example: the high proportion of elderly residents in the Tōhoku region left it more vulnerable to the ravages of the tsunami, as many lacked the ability to escape. As a result, the death toll amongst those aged 65 and above was four to six times higher than for other age groups, and in total accounted for more than 60 per cent of all tsunami deaths (Ando et al. 2013, pp. 2185–6). The greying of Tōhoku has increased at a faster pace following the disaster, as families and younger people have chosen to move elsewhere because of radiation fears and uncertainty over the future of the region. This creates difficult questions for rebuilding, as the long-term sustainability of these communities is increasingly doubtful, leaving those who remain – primarily the elderly – more vulnerable. From a human security perspective, the challenge is to find a way to rebuild these communities in a resilient and sustainable manner.

As a way of concluding, rather than recounting the contents of the preceding contributions, this chapter will consider a number of significant issues that have emerged out of them. While this book has focused on the case of the triple disaster that struck Japan in 2011, many of the issues discussed correspond to the human security problems that arise after any natural disaster (Hobson et al. 2014). There are two significant ways that this case differed from many others, however. First, as noted, it occurred in the context of an economically advanced, stable democracy, which provides an important counter to the assumption that human insecurity is only a problem for the developing world. Second, the nuclear accident at Fukushima – especially when combined with the tsunami – created a rather unique set of issues. Taken together, it is possible to see how, on the one hand, the human security approach helps shed light on this complex, multi-faceted disaster; and on the other, how this case assists us in further refining our understanding of human security.

Japan and human security

Despite the many examples of situations of extreme vulnerability generated by the triple disaster, and an explicit acknowledgement by the UN Secretary-General that this experience illustrated the relevance of the human security approach (UNSG 2012, p. 5), there are clear indications that within Japan it is still considered a concept relevant only for others. This could be seen at a 'High Level Symposium on Human Security', held as part of the Fifth Tokyo International Conference on African Development (TICAD) in June 2013. Admittedly the primary focus for this event was African development, but its remit did include conceptual discussion and reflection on the human security approach (MOFA 2013b). In his opening remarks, Prime Minister Shinzo Abe (2013) announced, 'Africa, as the land of hope, has a vast potential. I am convinced human security sheds a precious ray of light onto such [an] African future'. After highlighting some of the virtues of the human security approach – especially its dual concern with vulnerability and empowerment – he observed that 'these consequences of human security have many things in common with what I aim at through Japan's Growth Strategy' (Abe 2013). Abe did not consider his growth strategy as a way of further ensuring human security in Japan, but instead revealed the persistent assumption that this doctrine is only of relevance for the developing world. Not even Sadako Ogata (2013), former co-chair on the Commission on Human Security, referenced Japan in her talk. Despite one of the main concerns of the symposium being human security and health – an issue of great relevance in the Japanese context as the contributions from Hobson, Otani, and Sakamoto have shown – the triple disaster was not raised once throughout the event. While there has been an acknowledgement by Japanese scholars that human security is relevant at home (e.g. Osa 2012), it appears that Japanese policy-makers have yet to grasp this fundamental point.

Human security, as Bacon detailed in his chapter, has been seen as belonging exclusively to the preserve of Japan's Overseas Development Assistance (ODA). As the TICAD event indicated, this understanding continues to prevail. In Japan's official ODA policy the following elements are outlined as entailing a human security perspective:

1 Assistance that puts people at the centre of concerns and that effectively reaches the people;
2 Assistance to strengthen local communities;
3 Assistance that emphasizes empowering of people;
4 Assistance that emphasizes benefiting people who are exposed to threats;
5 Assistance that respects cultural diversity;
6 Cross-sectoral assistance that mobilizes a range of professional expertise.

(MOFA 2013a)

What should be clear from the preceding chapters is that these guidelines are equally relevant for responding to the 2011 earthquake, tsunami and nuclear accident. Considering that the rebuilding process to date has been plagued by delays, scandals and complaints about being overly centralized and bureaucratic, Japan's recovery would benefit from adopting the same principles it applies when providing assistance overseas. In this regard, Samuels (2013, Chapter 6) argues that the most significant change following the triple disaster may be the shift in power from the central government to local authorities in terms of disaster preparedness. This is a positive outcome from a human security perspective, and one that matches with the guidelines listed above.

When considering the relevance of the human security approach for Japan, it is especially revealing that one of two examples Japan's Ministry of Foreign Affairs (MOFA) provides of 'good practices on human security' in its ODA is a project assisting people in the Ukraine who were impacted by the Chernobyl nuclear accident. When the references to Chernobyl are removed, it becomes clear that the human security problems this project was meant to address are now equally present in Fukushima:

> The residents in the areas affected by the nuclear accident … are in need of psychological care as a result of the damage to their health and stress they suffered from the forced emigration. Moreover, the affected areas are also facing difficulties in economic recovery … and the effects of an aging of the population and depopulation. They are also struggling to overcome poverty.
> (MOFA 2013a)

MOFA (2013a) highlights how the project adopted a human security perspective through supporting community organizations that developed and executed their own recovery plans, which not only improved their living conditions, but also empowered people and helped them 'regain dignity as human beings'. If Japan has successfully executed a project that has helped to improve human security in areas affected by the Chernobyl nuclear accident, it should take that knowledge and directly apply it at home.

Acknowledging the relevance of human security for Japan could lead to a wider range of rebuilding strategies being considered, as other experiences would then be identified as relevant. For instance, in her chapter Fukushima considers how insights from peace-building can be applied in a disaster setting. Not only would this help with recovering from the triple disaster, it would also assist in Japan's preparations for future disasters. In this regard, a report by the Humanitarian Coalition (a group of major NGOs) on lessons learned from the disaster identified that its Japanese counterparts 'were not ready to deal with a disaster of this magnitude in their own country', and that Japan Platform, which was mandated to coordinate humanitarian assistance after the disaster, was not capable of doing so (Audet and Paquette 2012, pp. 5–8). Reflecting on this, Noriyuki Shiina, Japan Platform's Secretary-General, explained that, 'Japan Platform is designed to respond to emergency situations outside Japan. In 2006 we started to respond to disasters inside Japan, but we didn't have much experience with responding to domestic disasters' (IRIN 2012). A better appreciation that the same kind of humanitarian issues that Japan Platform and Japan's ODA address overseas can appear at home would make the country better prepared for future disasters. In this sense, an awareness of vulnerabilities can be empowering, by encouraging prevention and preparation.

Empowerment after the triple disaster

This volume has explored in detail the vulnerabilities that were created and exacerbated by the triple disaster. Contributors have also considered a range of ways that those affected have been able to maintain their agency during this difficult period. This reflects that human security is concerned not only with recognizing vulnerabilities; it also incorporates a strong emphasis on identifying how these can be combatted through empowering people. The significance of this more 'positive' dimension is noted by the Commission on Human Security:

> Empowerment is important because people develop their potential as individuals and as communities. ... People empowered can demand respect for their dignity when it is violated. They can create new opportunities for work and address many problems locally. And they can mobilize for the security of others ...
>
> (CHS 2003, p. 11)

This volume has explored these issues by emphasizing the social dimensions of human security. In so doing, it has sought to respond to Paul James's call for a radical rethink of human security, which entails developing a 'thicker' understanding of what 'human' means, and moving beyond conceiving of 'security' in purely military terms.

The approach to human security developed here has stressed the societal dimension of empowerment, by considering the role played by interpersonal bonds, communal relationships and cultural practices in rebuilding after the terrible events of 11 March 2011. Successful rebuilding is about much more than providing the necessary infrastructure; it is also about maintaining old social networks and building new ones, as Maly, Otani and Sakamoto all explore in their contributions. Meanwhile, Fukushima's chapter demonstrates how cultural practices specific to Tōhoku have helped people retain these communal ties, while also helping boost people's spirits and assisting in their recovery from deeply traumatic experiences. Another interesting example of this was Atsushi Chiba, an undertaker who volunteered to perform Buddhist rituals on nearly 1,000 bodies after the tsunami. Explaining his actions, Chiba stated that, 'yes, they are dead. But in Japan, we treat the dead with respect, as if they are still alive. It's a way to comfort the living' (Tabuchi 2012). Not only did his actions bring solace to the families of the deceased, his story was publicized through a popular book and movie, which in turn gave hope and courage to others. Understanding and incorporating such cultural awareness into disaster response and rebuilding is an important step from a human security perspective.

The disaster provided further evidence of Japan's strong social bonds and the sense of solidarity its people share. As Samuels (2013, p. 39) notes, victims of the triple disaster were 'widely admired – almost to the point of essentialist caricature – for their patient and persevering nature'. Even if this discourse was perhaps overplayed, there is certainly much truth in it and these social ties are important foundations for the rebuilding process. This point is made by Otani, who indicates that social capital is an important factor in the recovery process, an argument made in much greater detail by Aldrich (2012a). Applying his insights to the triple disaster, Aldrich (2012b, pp. 154–5) observes that 'individuals and localities [have] bounced back from tragedy and hardship not solely through wealth, government aid, or top-down leadership, but through their neighbors, connections and social networks'. The challenge is finding ways to maintain and support these bonds, especially now as they become frayed due to the protracted nature of people's displacement.

Concern and anger over the Fukushima Dai-ichi disaster has led to a considerable resurgence in civic activism. Protests have become far more common and much more diverse in their social and demographic composition. While anti-nuclear activists have failed to convince the Abe government, their protests have become too prominent to be comfortably ignored. Indeed, one might optimistically conclude that there has been a 'democratization' of energy policy in Japan, with it no longer being the exclusive preserve of the nuclear village. Some of the empowering aspects of this anti-nuclear sentiment were considered in the contribution from Bacon and Sato, who highlighted the unique deliberative polling exercise undertaken by the former DPJ government. This is a human security exercise *par excellence*, as it provides people with more information and encourages dialogue as a way of reaching a more considered decision on a certain policy, which has empowering consequences at both an individual and a community level. As Bacon and Sato suggest, there is much that could be gained from undertaking further deliberative polls.

One of the most significant 'positives' of the nuclear accident is a reconsideration in Japan of nuclear power and a much greater interest in developing renewable forms of energy, a topic also discussed by Bacon and Sato. There are too many developments and possibilities in Japan's renewables sector to adequately cover here (see DeWit 2012), but one example worth highlighting is the ambitious plans of Masayoshi Son, head of the telecommunications company Softbank, who has been investing heavily in renewable energies since 3/11. His goal is for renewables to account for 50–60 per cent of Japan's energy production by 2030, and he even aspires to construct an 'Asian Super Grid' that would connect Japan with electricity grids in South Korea, China and Mongolia (Obe 2012). Son's proactive response is illustrative of what a human security approach calls for. It is worth emphasizing that the push towards renewables is a positive development regardless of whether one is for or against nuclear power. By further developing renewables, Japan will have a more diverse energy portfolio, making it more resilient and prepared in the case of future energy crises, which remain a realistic possibility.

The jury is still out on how big an impact the triple disaster will have on Japanese politics and society, but the increase in civic activism and pressure for greater levels of transparency have the potential to foster a more engaged and responsive political system. As Samuels (2013, p. 200) concludes, as a result of the disaster, 'the rhetoric of crisis infused democratic politics, empowered new actors, stimulated long-awaited if piecemeal reforms, aroused considerable public protest, and may have pushed the policy process in the direction of transparency'. DeWit (2013) offers a more positive prognosis in observing that the reconstruction of the devastated Tōhoku region is

> blossoming into a clear vision, and an increasingly concrete reality, of ICT [information and communications technology] centred smart cities with resilient and efficient energy, water, transportation, communications, public safety, healthcare, and other social infrastructures.

Without denying the very serious human insecurities that remain, it is important to recognize these positive developments and investigate how they can be further built on.

Human security and the future

> The hardest thing is that we can no longer do the things that seemed so natural before.
>
> (Nobushi Kokubun, in Bird 2013)

While noting the way people have been empowered by the triple disaster, it is hard to escape the fact that for most affected people it has left them much more insecure. This difficult new reality is captured in this reflection from a Fukushima evacuee. Kokubun's observation also perfectly encapsulates what human security is ultimately about. As the UNDP report noted, 'most people instinctively understand what security means' (UNDP 1994, p. 3). It is that basic feeling of safety; of being free from severe and existential threats; and having a degree of confidence that this will continue into the future. In this regard, the Commission on Human Security notes, 'protecting a core of activities and abilities is essential for human security, but that alone is not enough' (CHS 2003, p. 4). Human security is meant to entail more than 'bare life'. It is also about providing the 'building blocks of survival, livelihood and dignity' (CHS 2003, p. 4). From this perspective, a central part of human security is protecting the 'pattern of our daily lives' and promoting the 'continuation of daily life' (UNDP 1994, p. 3; CHS 2003, p. 10). Notice the words used: 'pattern' and 'continuation'. Both indicate repetition, an assumption about the future. It is in this sense that human security is about creating conditions that allow for people to live their lives with some degree of confidence not only that they are safe today, but that they will remain safe tomorrow, the day after and the day after that. What a natural disaster does to those that survive it is disrupt, and in many cases destroy, this faith in the future.

When thinking about human security in terms of ensuring the 'continuation of daily life', it can be seen that the triple disaster has had far-reaching consequences, as the above quote from the Fukushima evacuee reminds us. On a very basic level it has disrupted daily life, and it has undermined confidence in the future for those whose lives were radically changed on 11 March 2011. The impact is most evident in the case of the nuclear accident. Evacuees have now been displaced for three years, many with no clear idea about when, if ever, they can return to their homes. Decontamination efforts have proven slower and less effective than the government hoped, extending the timeline for when people might be able to return. Reflecting on his plight, Ichiro Kazawa, an unemployed evacuee, observed, 'you can't have a temporary life forever … I think it will be easier for people who can't go back anyway to be told that so they can plan their future' (Knight and Slodkowski 2013). As a result of the nuclear accident, evacuees are left in limbo: unable to return to their old lives, but also unable to start new ones.

The impact of the nuclear accident on affected people's daily lives and their confidence in the future is particularly pronounced in regards to their health. In his report on the disaster, the United Nations Special Rapporteur on the right to health concluded that:

> The nuclear accident in Japan has affected the right to health of evacuees and residents alike and has had an impact on physical and mental health, particularly of pregnant women, older persons, and children. The precise health implications of radiation exposure are still not clear, as long-term health effects of low-dose ionising radiation are still being studied. The evacuation has caused the breakdown of families and communities, giving rise to mental health concerns, especially among first responders, older persons, mothers and children.
>
> (Grover 2013, p. 7)

Fear of radiation – an amorphous danger that you cannot see, touch or smell – is causing increased stress and anxiety. These are also caused by uncertainty about the future, with people fearful that they may become sick with leukaemia or cancer at some point in the future. Parents fear for the current and future health of their children. Nuclear workers worry about how their current employment may impact their future chances of having a family. Young women are concerned that the stigma of being from Fukushima will threaten their chances of finding a spouse. Fishermen wonder when, if ever, they will be able to catch fish to sell and not just for radiation monitoring. Farmers ask what they need to do to restore consumer trust in their produce. And, in turn, people across Japan remain doubtful about whether it is safe to eat produce or fish from near Fukushima.

It is not just at the individual level that uncertainty over the future exists. At the community level, as noted, there are major questions about whether it will be possible to rebuild many affected towns. Even the future of Fukushima Dai-ichi remains unknown: it is expected that it will take 30–40 years to complete the decommissioning process, but nobody knows for sure because nothing like this has ever been attempted before. Not even all the technology needed to complete the process has been built yet. At the societal level, there still remain huge questions about how to deal with contaminated areas, how to rebuild the Fukushima region, how to handle the truly colossal amount of debt from the nuclear accident. There are so many indeterminate issues that did not exist on 10 March 2011 but now make it difficult to have confidence in the future. The triple disaster, and especially the nuclear accident, has fundamentally shaken people's sense of well-being, and undermined the basic pattern of their daily lives. The great challenge that Japan faces is creating a new future for itself, and especially for the people that were affected by the triple disaster, one in which this confidence in the continuity of daily life is restored.

Preparing for the next disaster

The 2011 triple disaster was unprecedented in the country's history. The magnitude 9 quake was the largest ever recorded in Japan, and the fifth largest of the last century. The nuclear accident at Fukushima Dai-ichi was the second worst ever experienced. Despite Japan's long history with earthquakes and tsunamis, and it being regarded as a world leader in disaster preparation, the country was simply not ready for what happened on 11 March 2011. To quote former Prime Minister Kan: 'the cause of this catastrophe is, of course, the earthquake and the tsunami but, additionally, the fact that we were not prepared. We did not anticipate such a huge natural disaster could happen' (Biello 2013).

The Commission on Human Security, which was funded by the Japanese government, explained that: 'protecting people's security requires identifying and preparing for events that could have severe and widespread consequences' (CHS 2003, p. 10). Japan failed to do this in 2011. It is important that this does not happen again. There have been more than 2,800 earthquakes in Japan since 11 March 2011, and it appears that the Japanese archipelago is experiencing a period of increased seismic activity (Aulakh 2013; *Japan Today* 2013). Japan's Central Disaster Prevention Council recently estimated that if a magnitude 7 earthquake directly hit Tokyo, it would cause 23,000 deaths and ¥95.3 trillion in damage. And there is a 70 per cent chance of this occurring in the next 30 years. The report has been criticized for not examining the possibility of a greater magnitude quake, with suggestions that this is due to a desire to avoid concerns ahead of the 2020 Tokyo Olympics (*Japan Times*, 2013). Yet what the 2011 triple disaster demonstrated with horrible clarity is that preparations should be made based on the worst-case scenario. Such a disaster is liable, indeed likely to happen again, meaning that the best way that human security can be ensured is through prioritizing prevention.

Japan's human security challenges are closely connected to the future. It must find a way to rebuild the Tōhoku region and help in the restoration of people's daily lives, so there is a sense of certainty, regularity and continuity to them once again. It must also prepare itself for earthquakes, tsunamis and other possible disasters. Whether Japanese policy-makers realize it or not, human security *is* relevant for their country, and it is only through addressing these twin challenges that Japan will move closer towards a society in which 'all individuals, in particular vulnerable people, are entitled to freedom from fear and freedom from want, with an equal opportunity to enjoy all their rights and fully develop their human potential' (UNGA 2005, p. 31).

References

Abe, S. 2013, 'Opening remarks at symposium on human security', 2 June, http://www.mofa.go.jp/policy/oda/sector/security/pdfs/sympo1306_pm.pdf, viewed Dec. 2013.

Aldrich, D. 2012a, *Building Resilience: Social Capital in Post-Disaster Recovery*, Chicago, IL: University of Chicago Press.

Aldrich, D. 2012b, 'Social capital in post disaster recovery: Towards a resilient and compassionate East Asian community', in Y. Sawada and S. Oum (eds), *Economic and Welfare Impacts of Disasters in East Asia and Policy Responses*. ERIA Research Project Report 2011-8, Jakarta: ERIA, pp.153–74.

Ando, M., Ishida, M., Hayashi, Y., Mizuki, C., Nishikawa, Y. and Tu, Y. 2013, 'Interviewing insights regarding the fatalities inflicted by the 2011 Great East Japan Earthquake', *Natural Hazards and Earth System Sciences*, 13, pp. 2173–87.

Audet, F. and Paquette, F. 2012, *Lessons from the 2011 Japanese Earthquake and Tsunami for Canada's Humanitarian Sector*, report submitted to Humanitarian Coalition, 24 Sept., http://humanitariancoalition.ca/sites/default/files/publication/japan_en-29jan.pdf, viewed Jan. 2014.

Aulakh, R. 2013, 'Japan's tsunami: How clay on the Pacific floor could raise the risk of another large earthquake', *The Star*, 5 Dec., http://www.thestar.com/news/world/2013/12/05/japans_tsunami_how_clay_on_the_pacific_floor_could_raise_the_risk_of_another_large_earthquake.html, viewed Jan. 2014.

Biello, D. 2013, 'The nuclear odyssey of Naoto Kan, Japan's Prime Minister during Fukushima', *Scientific American*, 16 Oct., http://www.scientificamerican.com/article.cfm?id=nuclear-power-odyssey-of-naoto-kan-former-japan-prime-minister-during-fukushima, viewed Jan. 2014.

Bird, W. 2013, 'Two years on, Fukushima evacuees seek justice and a normal life', *Japan Times*, 10 March, http://www.japantimes.co.jp/life/2013/03/10/people/two-years-on-fukushima-evacuees-seek-justice-and-a-normal-life, viewed Dec. 2013.

Commission on Human Security. 2003, *Human Security Now*, New York: CHS, http://www.unocha.org/humansecurity/chs/finalreport/English/FinalReport.pdf, viewed Aug. 2013.

DeWit, A. 2012, 'Japan's remarkable renewable energy drive: After Fukushima', *Asia-Pacific Journal*, 10/11(10), http://www.japanfocus.org/-Andrew-DeWit/3721.

DeWit, A. 2013, 'Just gas? Smart power and Koizumi's anti-nuclear challenge', *Asia-Pacific Journal*, 11/50(3), http://www.japanfocus.org/-Andrew-DeWit/4049.

Grover, A. 2013, *Report of the Special Rapporteur on the Right of Everyone to the Enjoyment of the Highest Attainable Standard of Physical and Mental Health. Addendum: Mission to Japan*, Human Rights Council, 23rd session, A/HRC/23/41/Add.3, 2 May, http://www.ohchr.org/Documents/HRBodies/HRCouncil/RegularSession/Session23/A-HRC-23-41-Add3_en.pdf, viewed Dec. 2013.

Hobson, C., Bacon, P. and Cameron, R., eds. 2014, *Human Security and Natural Disasters*, London: Routledge.

IRIN. 2012, 'Disasters: Learning from Japan's tsunami', IRIN humanitarian news and analysis, 9 March, http://www.irinnews.org/report/95039/disasters-learning-from-japan-s-tsunami, viewed Jan. 2014.

Japan Today 2013, 'Stepped-up seismic activity raises concerns that big quake may be looming', 6 Feb., http://www.japantoday.com/category/kuchikomi/view/stepped-up-seismic-activity-raises-concerns-that-big-quake-may-be-looming, viewed Jan. 2014.

Japan Times 2013 'Critics blast quake plan as too soft', 20 Dec., http://www.japantimes.co.jp/news/2013/12/20/national/critics-blast-quake-plan-as-too-soft, viewed Jan. 2014.

Knight, S. and Slodkowski, A. 2013, 'For many Fukushima evacuees, the truth is they won't be going home', *Reuters*, 11 Nov., http://www.reuters.com/article/2013/11/11/us-japan-fukushima-idUSBRE9AA03Z20131111, viewed Jan. 2014.

Ministry of Foreign Affairs Japan. 2013a, 'Human security: Japan's action', MOFA, http://www.mofa.go.jp/policy/oda/sector/security/action.html, viewed Dec. 2013.

Ministry of Foreign Affairs Japan 2013b, 'Symposium on human security: Overview and evaluation', MOFA, 3 June, http://www.mofa.go.jp/policy/oda/page_000006.html, viewed Dec. 2013.

Obe, M. (2012) 'First the iPhone: Now renewables', *Wall Street Journal*, 18 June, http://online.wsj.com/news/articles/SB10001424052702304371504577404343259051300, viewed Dec. 2013.

Ogata, S. 2013, 'Development of the concept of human security', 2 June, http://www.mofa.go.jp/policy/oda/sector/security/pdfs/sympo1306_ogata.pdf, viewed Dec. 2013.

Osa, Y. 2012, 'Human security vs. national security in a developed country: Japan's triple disaster and the challenges of Japanese civil society', presentation at 'Human Security and Natural Disasters' workshop, United Nations University, Tokyo, 21 Feb.

Samuels, R. J. 2013, *3.11: Disaster and Change in Japan*, Ithaca, NY: Cornell University Press.

Tabuchi, H. 2012, 'Japan finds story of hope in undertaker who offered calm amid disaster', *New York Times*, 10 March, http://www.nytimes.com/2012/03/11/world/asia/a-year-later-undertakers-story-offers-japan-hope.html, viewed Dec. 2013.

UNDP. 1994, *Human Development Report 1994*, New York: Oxford University Press.

United Nations General Assembly. 2005, *2005 World Summit Outcome*, UNGA A/RES/60/1, http://www.un.org/womenwatch/ods/A-RES-60-1-E.pdf, viewed Jan. 2014.

UNSG. 2012, *Follow-up to General Assembly Resolution 64/291 on Human Security: Report of the Secretary-General*, UNSG A/66/763, 5 April, https://docs.unocha.org/sites/dms/HSU/Publications%20and%20Products/Reports%20of%20the%20Secretary%20General/A-66-763%20English.pdf, viewed 27 July 2013.

Index

Page numbers followed by *n* indicate note numbers

CPSIA information can be obtained
at www.ICGtesting.com
Printed in the USA
BVOW06s0351160817
492160BV00006B/51/P